WOMEN AND PHILANTHROPY IN NINETEENTH-CENTURY ENGLAND

Caricatures of district visitors, from *A Parson, My District Visitors*, 1891

WOMEN AND PHILANTHROPY IN NINETEENTH-CENTURY ENGLAND

by

F. K. PROCHASKA

CLARENDON PRESS OXFORD
1980

Oxford University Press, Walton Street, Oxford OX2 6DP

OXFORD LONDON GLASGOW
NEW YORK TORONTO MELBOURNE WELLINGTON
KUALA LUMPUR SINGAPORE JAKARTA HONG KONG TOKYO
DELHI BOMBAY CALCUTTA MADRAS KARACHI
NAIROBI DAR ES SALAAM CAPE TOWN

Published in the United States
by Oxford University Press, New York

British Library Cataloguing in Publication Data

Prochaska, F. K.
Women and philanthropy in nineteenth-century England
1. Woman philanthropists – England – History – 19th century
I. Title
361.7 HV245 80-40192

ISBN 0-19-822627-6
ISBN 0-19-822628-4 Pbk

Typeset by Hope Services, Abingdon
Printed in Great Britain by
Lowe and Brydone Ltd., Thetford

For
Elizabeth and Franklin,
Sheila and John

Preface

My research springs from an interest in philanthropy rather than in feminism. The original object in undertaking this study was to fill one or two gaps in David Owen's distinguished *English Philanthropy, 1660–1960*, but I would like to think that it has developed into a contribution to the history of the emancipation of women as well. Most of the topics that appear were not anticipated when I began this project. I was surprised, for instance, to find the fund-raising activities of children, mostly little girls, to be so extensive. Nor had I realized just how important the charity bazaar was to the expansion of philanthropic enterprise. Most of all, my researches brought home to me the inescapable importance of religion in the lives of nineteenth-century women. In other words, the chapter headings suggested themselves from the recurring themes in the evidence rather than from any preconceived ideas.

While I was happy to allow the issues to emerge from the sources, I should say that my initial intention was to avoid writing a series of biographical sketches or purely institutional studies. I hoped to find topics which cut across these methods, which would suggest the variety and scope of female benevolence while contributing to the study of nineteenth-century opinion. I confess that the typical has interested me rather more than the exceptional, and I have consciously given as much attention to little-known women and obscure charities as to the well established. This will alarm some readers who would like to see more space devoted to, for example, Florence Nightingale and the Charity Organisation Society. In my defence I would say that in a study as broad as this much has had to be left out, and I would add that there is little point in telling people what they already know.

I have included information on working-class charity in the text, but a history of women and philanthropy, which emphasizes organized charity, is not the best place to discuss this subject at length. As the philanthropy of the poor was usually informal and undocumented it makes any systematic analysis of their benevolence difficult. We should, however, keep in mind the view of a London cleric, who remarked, in words reminiscent of Engels, that 'the poor breathe an atmosphere of charity. They cannot understand life without it. And it is largely this kindness of the poor to the poor which stands between our present civilisation and revolution.'[1]

This work was liberally supported by a Fellowship from the American Council of Learned Societies and a Research Council Grant from the University of Missouri at Columbia, where I was a visiting Assistant Professor in 1973–4. I would like to thank them and also the editors of the *International Review of Social History*, the *Journal of British Studies*, and *The Journal of Imperial and Commonwealth History*, who have given me permission to republish material which appeared in their journals in an earlier form. The staff of the British Library, the Colindale Newspaper Library, the Public Record Office, the Greater London Record Office, the Bodleian Library, the University of London Library, the School of Oriental and African Studies Library, the University of Missouri Library, the University of Wisconsin Library, the former Fawcett Library, the Friends Library, the Wellcome Library, and other institutions, among them many charitable societies, have given me the most painstaking assistance. I should like to single out the Archivist of the British and Foreign Bible Society, Miss Kathleen Cann, for going beyond the call of duty in unearthing various letters and pamphlets which I would have missed if left to my own devices.

Friends and colleagues in England and America have given generously of their time and I fear that, on occasion, I have tried their patience with my preoccupations. I would especially like to thank the staff of the Institute of Historical Research, University of London, and the members of the Institute Seminar run by Professor I. R. Christie and Dr John Dinwiddy, who have taken the trouble to discuss various

[1] William Conybeare, *Charity of Poor to Poor, facts collected in South London at the suggestion of the Bishop of Southwark*, London, 1908, p. 6. See Friedrich Engels, *The Condition of the Working Class in England*, eds. W. O. Henderson and W. H. Chaloner, Stanford, 1958, pp. 102, 140.

papers of mine over the past few years. Michael Collinge has been a knowledgeable guide to sources and read the manuscript in its entirety, and I am under a great debt to him for his encouragement and criticism. I am also indebted to Claire Barwell, Ivon Asquith, and the expert readers of the Oxford University Press, who coupled their valuable comments with lessons on the art of erasure. Others have kindly given me assistance on detailed points, among them Sue Brown, Deborah Cherry, Elizabeth Danbury, Janice Hadlow, Michael Harris, Richard Hunter, William Kellaway, Bob Ruigh, Bernard Semmel, Rosemary Taylor, Rachel Vorspan, Diane Worzala, and Robert and Susan Youngs; and others still have contributed unknowingly. My greatest debt is to my wife and fellow historian, Alice Prochaska, who first suggested that I work on nineteenth-century philanthropy and who has read and read again the drafts of this book with charity and sagacity. She will forgive me for dedicating it to others, close to us both.

London, 1979

Table of Contents

List of Illustrations

The Illustrations have been reproduced by permission of the British Library.

INTRODUCTION

Woman's 'Nature and Mission'

English writers of the nineteenth century often used the phrase woman's 'nature and mission' when they wished to discuss what they believed to be the characteristics of women and the functions which society encouraged them to undertake. There were, of course, as many missions as there were women in nineteenth-century England, for each woman, however humble, must have had a notion of her uniqueness and of the individuality of her circumstances and day-to-day tasks. This is not to say that she was free to go her own way unencumbered by law and custom; society imposed severe limitations on women, some of which have persisted. Yet one is struck by the way so many women, and not only of the middle and upper classes, led independent lives despite the weight of convention. There were pressures on them 'to suffer, and be still', as the writer and philanthropist Sarah Ellis put it, but this should not lead us to conclude that they necessarily did so.[1] We are perhaps too prone to see limitations where the women of the past saw possibilities. The closer we come to their lives the more variety we find.

One thing is certain: the view of woman's mission was in flux in the nineteenth century. As the conditions of daily life changed the gap widened between what the conventional wisdom expected of women and what was their real status. Industrial and demographic change was working to complicate and reshape the relations between the sexes. Life in cities, for instance, was less domestic than in the country, and social life generally was becoming more and more public.[2]

[1] Mrs Ellis, *The Daughters of England*, London [1845], p. 161. See also *Suffer and be Still*, ed. Martha Vicinus, Bloomington and London, 1972, p. x.

[2] *Woman's Mission*, ed. Baroness Burdett-Coutts, London, 1893, pp. xiv–xv; Georgiana Hill, *Women in English Life from Medieval to Modern Times*, 2 vols., London, 1896, ii, p. 100.

Working-class women joined industry in substantial numbers; middle-class women, with greater wealth, found they had more time on their hands. In their different ways, both groups were being thrown back on their own resources. In the case of the large number of 'surplus women' without husbands, a problem highlighted by the popularization of the mid-century censuses, the law of necessity created its own demands, which called into question traditional views about women's position in society. In short, the changing facts of daily existence encouraged, often required, women to alter and to expand their routine. As their sphere widened so too did their sense of mission, which in turn made further inroads into the conventional attitudes towards women.[3]

Take an example of this process. In the early nineteenth century it was virtually unheard of for a woman to make a public speech. In some circles men discouraged them from attending public meetings, though on occasion they slipped in unnoticed, to sit demurely behind the organ screen.[4] Things had changed dramatically by the early twentieth century; a woman no longer handed over her manuscript to be read out aloud by a man; she no longer blushed at the prospect of mounting a platform. When suffragettes chained themselves to the railings outside the Houses of Parliament, public speaking did not seem such an affront to humanity. (Dr Johnson's dog had found its footing, so to speak.) Such changes came about only gradually, touched off by women determined to get their message across and willing to test convention by addressing charity meetings, social science congresses, and trade union gatherings. By such actions they broke down the prejudice against women speakers and made it easier for the less forthright to express themselves in public without fear of obloquy. By enlarging the scope of women's activities, they also modified the way in which people interpreted the possibilities inherent in the female character.

What were traditionally thought to be a woman's special

[3] Hill, *Women in English Life*, ii, pp. 88-92.

[4] At a meeting of the Society for the Propagation of the Gospel in the 1820s, several ladies were hidden behind the organ. 'A bishop was publicly rebuked by a baron of the exchequer for bringing in his own wife upon his arm.' *The Christian Observer*, Jan. 1861, p. 40.

traits? Answers to this absorbing question were not simply descriptive, but sought to arouse women to live up to their calling and to deter them from what were thought to be their besetting sins: idleness and pretension. Moral, modest, attentive, intuitive, humble, gentle, patient, sensitive, perceptive, compassionate, self-sacrificing, tactful, deductive, practical, religious, benevolent, instinctive, and mild; these and their synonyms were used over and over again by writers, male and female, both opposed to and in favour of the emancipation of women, who sought to fathom or to shape women's behaviour. Most of these words were flattering, yet they could be double-edged, for if they were earned it was often at the expense of a more rounded existence. Dealing largely with the affections, they tended to be thought of as virtues associated with family life, and could be obstacles in the more competitive society outside the home. But this was not a very great worry to most writers. They commonly accepted the doctrine of the two spheres, whose most eloquent exposition was in Ruskin's *Sesame and Lilies*: women were to provide that sympathy and 'sweet ordering', which complemented man's capacity to create and to rule. 'The feminine principle in human nature', remarked one female writer, '. . . taking the hand of the masculine principle, helps on the work of England's present new life to its perfection'.[5]

Most parents were anxious to encourage the traits believed to be feminine in the education of their daughters. The variations in female education were vast, ranging from virtually none at all, through what might be picked up in the home or at a Dame, Industrial, or Sunday school, to the rigours of academic life at an institution like Cheltenham Ladies' College, opened in 1854. But whatever the opportunities, it was an education usually different in kind from what a boy of equivalent class might expect. Firstly, religion was more pronounced in the training of girls. Secondly, the world of girls was an indoor world, for outdoor games and exercise rarely played such a prominent part in their activities. As one 'child of the seventies' put it, 'my outside amusements were mainly pale reflections of what the boys told me about

[5] Frederika Bremer, *England in 1851; or, Sketches of a Tour in England*, Boulogne, 1853, p. 6.

theirs'.[6] Thirdly, but not least in importance, girls were very likely to be trained in the care of households. The dolls and needles surrounding a middle-class girl would have made little impression on a brother who might be reading Tacitus or at least romping in the open air. His education tended to be challenging to both mind and body; hers, whether under the watchful eye of a governess or not, dwelt on domestic concerns and the cultivation of the sentiments. 'Love of intellectual pleasures is as fatal a snare as the love of sensual pleasures', remarked one female authority on the schooling of girls.[7] Parents and teachers were to enhance the virtues inherent in female nature, not lead them down the path to worldliness and vice.

Despite its failings, which reformers like Emily Davies began to point out in the mid-Victorian years,[8] such an education had its rationale and a very great number of supporters in the nineteenth century. Did it not lead to domestic tranquillity, they argued, the promotion of a more refined society, and the protection of public morals? Was it not true to all that was best in the female sex? Sanctioned by nature, did it not also fulfil the biblical imperative: woman is the helpmeet of man? And did it not follow, they might have added, that such training made sense in a world in which most women aspired to marry and in which the care of children was likely to take up their most productive years. The cycle of life told very heavily on women; their lives passed without much interruption from infancy and childhood to the rearing of their own children. In the mid-Victorian years married women each underwent, on average, six confinements.[9] The nursery and the sick chamber, the home and the neighbourhood, the church and the foot of the cross were the anchors of a woman's existence and excellence.[10] Of what use was an education that led elsewhere?

[6] M. Vivian Hughes, *A London Child of the Seventies*, London, 1934, p. 34.

[7] M. A. Stodart, *Hints on Reading; addressed to a young Lady*, London, 1839, p. 50. See also, A Lady, *Woman! as Virgin, Wife and Mother*, London [1838]; Thomas Broadhurst, *Advice to Young Ladies on the Improvement of the Mind and the Conduct of Life*, London [1808].

[8] See Sara Delamont, 'The Contradictions in Ladies' Education', *The Nineteenth-Century Woman: Her Cultural and Physical World*, eds. Sara Delamont and Lorna Duffin, London 1978, pp. 134–63.

[9] J. A. Banks, *Prosperity and Parenthood*, London, 1954, p. 3.

[10] *Blackwood's Magazine*, vol. liv, 1843, p. 397.

The traditional views on female education adjusted only slowly to the social change wrought by industrial and commercial expansion. In the lives of many women this expansion brought increased wealth, fewer occupations, and a greater reliance on servants. Domestics made up an increasing share of the labour force until late in the century,[11] and their employers were to be found quite low down on the social scale, particularly after 1860.[12] Freed from many of their domestic duties, women commonly found that there were few alternatives to lives of refined idleness. Denied specialized training in an age that increasingly demanded it, they found it difficult to compete with better qualified men in those trades and businesses that traditionally were open to them.[13] At the same time, added income made more and more women, the wives of tradesmen for example, aspire to a life of less domestic drudgery and more freedom.

During much of the nineteenth century there was very little employment suited to middle-class women outside writing and governessing, and many people of a refined sensibility thought the latter employment gave a lady rather too much of an air of necessity. It could be said that 'a lady, to be such, must be a mere lady, and nothing else. She must not work for profit, or engage in any occupation that money can command, lest she invade the rights of the working classes.'[14] As a letter to *The Times* put it in 1801: 'She cannot work–she cannot beg.'[15] Her frequent complaint was of having nothing to do. Some complained that she gave in too easily to dissipation and the fine arts. And some, very sensibly, offered her new possibilities.

Philanthropy was the vocation that most often sprang to mind. Throughout the nineteenth century it was seen as the leisured woman's most obvious outlet for self-expression. Benevolence, wrote the learned Catharine Macaulay, was one of the most 'animating of the moral principles', and if

[11] Phyllis Deane and W. A. Cole, *British Economic Growth, 1688–1959*, Cambridge, 1962, Table 30, p. 142.

[12] See page 151.

[13] See Ivy Pinchbeck, *Women Workers and the Industrial Revolution, 1750–1850*, London, 1930, p. 304.

[14] From Margaretta Greg's Diary, quoted ibid., p. 315.

[15] 26 December 1801.

genuinely practised it would 'entirely subdue the daemon *Ennui*'.[16] Hannah More, the most influential female philanthropist of her day, and probably the most influential woman of her day, spoke for many of her sex when she said that 'the time and money . . . snatched from vain and frivolous purposes, are more wisely directed together into the same right channel of Christian Benevolence'.[17] 'In charity', said a writer in *The English Woman's Journal* years later, 'there will ever be found a congenial sphere for the fruition of the unemployed energies of women'.[18] Whether casual or institutional, charitable work was relatively free from the restraints and prejudices associated with women in paid employments. Nor did the law have to be changed to permit women to expend their benevolent energy. 'As society is at present constituted', remarked Sarah Ellis in 1869, 'a lady may do almost anything from motives of charity or zeal, . . . but so soon as a woman begins to receive money, however great her need, . . . the heroine is transformed into a tradeswoman'.[19] The limitations imposed upon women combined with their training and experience to push more and more of them into voluntary work.

Active benevolence was most compelling to leisured women who were 'compassionate' and 'self-sacrificing' and who were traditionally skilled in caring for the young, the sick, the elderly, and the poor. And it did not need to be full-time work, for it could be fitted into household routine. Many writers argued that it trained women to be better wives and mothers; and those who never married need not be disheartened either, for their status was a 'providential arrangement adapted to the condition and wants of the world'. Spinsters, so the argument continued, were to 'feel themselves married to every creature of the race'.[20] There was, to be sure, enough poverty and distress in English

[16] Catharine Macaulay Graham, *Letters on Education*, London, 1790, p. 290.

[17] Hannah More, *Moral Sketches of Prevailing Opinions and Manners*, London, 1819, p. 214.

[18] *The English Woman's Journal*, iii, 1 May 1859, p. 196.

[19] Mrs Ellis, *Education of the Heart: woman's best work*, London, 1869, p. 14.

[20] Willian Landels, *Woman's Sphere and Work, considered in the light of Scripture*, London, 1859, pp. 142–4.

society to satisfy every woman's charitable inclinations, whatever her marital status. Nor did it hurt her cause that indigence was seen as the result of moral failing, for the religious and domestic skills of women were ideally suited to the moral reformation thought to be the best remedy for indigence.

Free from the cut and thrust of commercial life and thought to be more sensitive to personal relations, women were increasingly called upon to be agents of social improvement. Blessed with what has been called the 'non-invidious' temperament, they could claim a sense of community not so often found in men.[21] As one woman writer argued: 'Men, engaged in the active affairs of life, have neither time nor opportunity for those innumerable little acts of consideration which come within the sphere of female duty, nor are they by nature so fitted as woman for entering into the peculiarities of personal feeling, so as to enable them to sympathize with the suffering or the distressed.'[22] From their domestic citadel, women could make forays to spread that tenderness and purity, thought to be the essence of female character, through society.[23]

A distinctive feature of women's work in nineteenth-century philanthropy is the degree to which they applied their domestic experience and education, the concerns of family and relations, to the world outside the home. The saying 'charity begins at home' had a meaning unsuspected by its originator, for that was precisely where so many women developed the sympathies and skills necessary to perform good works in a wider sphere. As one lady phrased it, woman's missionary spirit was simply 'the flow of maternal love'.[24] Reformers from Hannah More to Millicent Fawcett argued that morality, self-denial, and compassion, woman's domestic virtues, were just what was needed in English public

[21] Thorstein Veblen, *Theory of the Leisure Class*, London, 1924, pp. 332-62. See also Ronald G. Walton, *Women in Social Work*, London, 1975, p. 85.

[22] [Sarah Ellis], *The Wives of England; their relative duties, domestic influence, and social obligations*, London [1843], pp. 294-5.

[23] A Woman, *Woman's Rights and Duties, considered with relation to their Influence on Society and on her own Condition*, London, 2 vols., 1840, *passim*. See also, *The Edinburgh Review*, lxxiii, April, 1841, pp. 189-209.

[24] A Woman [Sarah Lewis], *Woman's Mission*, London 1839, p. 128.

life. 'We want her sense of the law of love to complete man's sense of the law of justice', wrote the philanthropist and suffragist Frances Power Cobbe, 'we want her influence, inspiring virtue by gentle promptings from within to complete man's external legislation of morality. And, then, we want woman's practical service. We want her genius for detail, her tenderness for age and suffering, her comprehension of the wants of childhood to complete man's gigantic charities.'[25] As the family received more and more attention as the paramount social unit in society, women, believed to be the natural guardians of the household, gained new confidence. In their attempts to extend their influence they willingly reinforced the stereotypes of women as the more compassionate, self-sacrificing sex. The claims of women to moral authority and greater social recognition depended on public belief in their special and essential qualities.[26]

If it was a small step from the love of family to the love of the family of man, it was a step made easier by Christian teaching. All Christian denominations stressed the importance of charitable conduct, but none gave greater emphasis to it than the evangelicals, whose power was in the ascendancy from the late eighteenth century. The enormous influence of the evangelicals, including here all Bible Christians, was not unconnected with their individualist ethic: that each person worked out his or her own salvation was very much in accord with the *laissez-faire* ethos of the secular world. By the nineteenth century, doctrinal differences between Methodists and Church of England evangelicals had been subordinated; the Calvinist strait-jacket had been discarded. As William Wilberforce remarked, 'deliverance is *not forced upon us*, but *offered*

[25] Frances Power Cobbe, *Essays on the Pursuits of Women*, London, 1863, p. 32. See Mrs H. Fawcett, *Home and Politics*, London, 1898; *Macmillan's Magazine*, v, 1861, p. 92.

[26] Françoise Basch, *Relative Creatures: Victorian Women in Society and the Novel, 1837–67*, London, 1974, p. 269. For an interesting discussion of this theme in the American context see Kathryn Kish Sklar, *Catherine Beecher: A Study in American Domesticity*, New Haven and London, 1973, *passim*, and Nancy F. Cott, *The Bonds of Womanhood, "Woman's Sphere" in New England, 1780–1835*, New Haven and London, 1977, pp. 197–206.

to us'.[27] John Wesley would have agreed. With grace conditional and backsliding a pitfall there was a new emphasis given to good works. To evangelicals, as Wilberforce put it, the love of one's fellow man was 'the indispensable, and indeed the characteristic duty of Christians'.[28] It could be a matter of life and death. Citing Matthew 25: 36, Wesley warned his followers that they must assist the poor if they wished 'to escape everlasting fire, and to inherit the everlasting kingdom'.[29] And to Bible Christians 'everlasting fire' was no metaphor.

As a religion of duty, which placed service above doctrine, evangelicalism appealed particularly to women. Religious sensibility and social pity stood much higher in their minds than abstract, frequently arid theology. And if right conduct and moral fervour were thought to be their preserve, it would only be natural for evangelicalism, which put such a premium on manners and morals, to find a refuge in the female breast. As Wilberforce argued in *A Practical View*, after the Bible perhaps the most influential book of its day, women were more favourably disposed to religion and to good works than men; and he advised them to measure their spiritual progress by their improvement 'in love to God and man'.[30] Wesley reminded women of Phoebe's work in the early Church and concluded: 'Whenever you have opportunity, do all the good you can, particularly to your poor, sick neighbour. And every one of *you* likewise "shall receive *your* own reward, according to *your* own labour".'[31] Countless women writers throughout the nineteenth century repeated and elaborated such views in their novels, tracts, hymns, memoirs, and poems. Take, for example, the simple religious doctrine which shines through the lines of Martha Kenney's poem 'Charity', written on behalf of the Church Missionary Society in 1823:

[27] William Wilberforce, *A Practical View of the Prevailing Religious System of Professed Christians in the Higher and Middle Classes in this Country contrasted with real Christianity*, London, 1797, p. 50. See also Bernard Semmel, *The Methodist Revolution*, London, 1974, pp. 106–9. There is a good discussion of the relation of charity to evangelical doctrine in Elizabeth Jane Whately, *Evangelical Teaching; its meaning and application*, London, 1871.

[28] Wilberforce, *A Practical View*, p. 336.

[29] *The Works of the Rev. John Wesley*, 14 vols., London, 1872, vii, p. 123.

[30] Wilberforce, *A Practical View*, p. 434, 448.

[31] *The Works of the Rev. John Wesley*, vii, p. 126.

Oh Charity! thou fairest birth
Of heavenly Mercy! Sent on earth
To heal the wounds that sin has given,
To point the Christian's path to Heaven.[32]

Such literary effusions complemented the rigorous religious education commonly given to the female sex. Church and chapel going, Scripture readings and family prayers, working parties and mothers' meetings, and visiting the homes of the poor were habits common to millions of nineteenth-century families; and they often brought girls into contact with their needy neighbours. This contact was perhaps greatest among evangelicals, for women were rather more prominent in church and chapel activities in evangelical circles. The female leaders of Methodist classes and women Quaker ministers, among others, probably opened up spheres of charitable work through their various religious and social functions. And the more these functions took place at home the more likely they were to be dominated by wives and mothers. Women who wanted to expand their social contacts and vary their weekly routine must have seen an opportunity in any informality in church or chapel activities. These activities, in turn, often became formal and ritualized in the home. William Hale White, 'Mark Rutherford', who reluctantly attended the Dorcas meetings organized by the ladies of his congregation, ostensibly to make clothes for the poor, remembered them as oppressive occasions, with the sectarian gossip and religious 'slip-slop' only relieved by tea and buttered toast.[33]

The formality and enforced idleness of middle-class family life exasperated girls anxious to get out and be of use in the world. Beatrice Potter (later Webb) complained of the code of feminine domesticity that acted as a brake on her wider ambitions. The pressure was such that she 'silently withdrew' her plans for social work from family discussion.[34]

[32] Martha Kenney, *Charity: a poem*, London, 1823.

[33] *The Autobiography of Mark Rutherford*, ed. Reuben Shapcott, London, 1881, pp. 37–41. In Candleford Green such 'working parties' were also common and, though useful, were known also as 'clearing houses for gossip'. See Flora Thompson, *Lark Rise to Candleford*, London, 1945, pp. 447–8.

[34] Beatrice Webb, *My Apprenticeship*, London, 1926, pp. 116–17.

The endless round of entertaining and being entertained, of stitching and serving, could be so restrictive that it impelled imaginative and energetic girls to seek a wider field for the expression of their talents. Florence Nightingale, for one, felt an enormous sense of release when she finally left home to pursue a career in nursing.[35] Escape from the family circle was not always easy; among the most common routes were the death of a parent or a want of suitors. In the 'intellectual aristocracy' things could be easier, for here the girls were better educated and were more likely to be encouraged to take up some form of social service.[36] For others, charitable work, free from chaperons and prying relatives, represented deliverance from the stitch-stitch-church-stitch routine of female existence. It was adventure.

In many families restrictions on female dress and entertainments reduced the diversions which might stand in the way of practical service. Evangelicals generally and Quakers in particular were shut out from many of the ordinary sources of emotion, and benevolence became more important to them than it might have been otherwise. Both the activities and restrictions of nineteenth-century family life and female education tended to focus the affections and raised philanthropy to the level of obedience to God. Obedience to God necessitated the love of one's fellow man. As every student of the New Testament knew, the word charity itself was Greek for love, synonymous with Christ-like conduct. Whether evangelical, High Church, Catholic, or Unitarian, all agreed: to be Christian was to be charitable, or to quote one female writer, 'uncharitableness is that which strikes at the foundations of Christianity'.[37] Raised in a Christian society and believing themselves more compassionate than men, women willingly accepted charitable work as their rightful mission. The very word mission, used repeatedly and advisedly, was explicit.

There were reasons why women might have thought

[35] Cecil Woodham Smith, *Florence Nightingale*, London, 1950, pp. 84–116.

[36] See N. G. Annan, 'The Intellectual Aristocracy', *Studies in Social History*, ed. J. H. Plumb, London, 1955, p. 251.

[37] A Lady, *The Whole Duty of Woman or, a Guide to the female sex, from the age of sixteen to sixty*, Stourbridge, 1815, p. 23.

Christianity restrictive, yet most female reformers hailed it as an emancipating influence. It was more than a treasure chamber of consolations, though it was often that, for it heightened women's self-esteem and gave them a sense of place and direction. 'You have been dowered with Christianity', exclaimed the writer and rescue worker Ellice Hopkins, 'the one great historical religion that enshrines the sanctity of your womanhood'.[38] She and many others argued that as Christianity advanced the inequalities between the sexes would disappear. Not only the biblical scholars had heard of the saying, attributed to Christ by Clement of Alexandria, that 'the kingdom of God can only come when "two shall be one, and the man as the woman" '.[39] Such a statement reinforced women's liberal reading of the Gospels. 'It is impossible to dwell too forcibly upon the fact', remarked one essayist, 'that a diffusion of the Christian ideas and spirit in Society is the only safeguard of the rights of woman'.[40] 'Christianity', pronounced the female author of *Woman's Rights and Duties*, 'has done more to raise and uphold the condition of women than any other cause'.[41]

Among its many merits, Christianity was thought to give enormous scope to woman as wife and mother and by extension, enormous benefits to society. 'It enters the heart and love springs up at its appearing, it enters the house and *peace* is its attendant; wherever it is, whoever it influences, order, self-denial, activity, and benevolence are its attendants.'[42] But Christianity did not restrict woman to the role of domestic helpmeet; there was God's work to be done outside the home as well. 'Alive to every tender feeling, to deeds of mercy ever prone',[43] she should also, as the Bible enjoined, 'stretcheth

[38] Ellice Hopkins, *Grave Moral Questions*, London, 1882, p. 66. Proponents of natural religion appeared to have similar views. One of them wrote 'were a bridge thrown across our Channel, the whole [female] sex would be seen rushing to the British shores'. Anon., *Woman: as she is, and as she should be*, 2 vols., London, 1835, i, p. 3.

[39] *The Westminster Review*, cxxvii, 1887, p. 72, note.

[40] *North British Review*, xiv, 1851, p. 277.

[41] A Woman, *Woman's Rights and Duties*, i, p. 177. One woman's analysis of the Bible led her to conclude that Christianity justified equal pay for women and their right to sit on all public bodies on equal terms with men. L. Sapsworth, *The Emancipation of Woman*, London, 1913, pp. 16–17.

[42] Clara Lucas Balfour, *A Sketch of Mrs. Ann H. Judson*, London, 1854, p. 4.

[43] 'Praise of Woman', a poem by Mrs Letitia Barbauld, quoted in Jabez Burns, *The Marriage Gift Book and Bridal Token*, London, 1863, p. 69.

out her hand to the poor; yea, . . . reacheth forth her hands to the needy' (Proverbs 31: 20). Such passages of Scripture encouraged her to break out of her domestic circle and diffuse those ornaments of her nature: 'that warm sensibility of heart, that gentle, modest, and retiring delicacy of feeling, that disinterested, ardent, and enthusiastic affection'.[44] The Bible gave women opportunities as well as demanding sacrifices. Steeped in Scripture, women proclaimed for themselves 'no less an office than that of instruments (under God) for the regeneration of the world'.[45]

The influence of the Bible on the female mind is a common theme in the hundreds of nineteenth-century memoirs, diaries, and autobiographies that women have left us.[46] These sources are very often little more than 'devotional remains', chronicles of family life and religious experience. Recurring subjects include visiting the poor, Sunday school teaching, mothers' meetings, sewing garments for the poor, and a variety of other customary practices centered on the home and the church. Such matters may appear humdrum and stark, but for the fervent Christian the routine of domestic labour and charitable duty had its rationale and its rewards. Most of the writings are hemmed in by literary conventions, perhaps dictated by the religious publishers who brought out many of the works, but more likely dictated by the way in which the women saw themselves. It would be rash to argue that the lives of these women are distorted because of these literary conventions. The wives and daughters of ministers wrote many of the memoirs, but a large number were written by women distinguished only by a desire to tell a story of conversion and service. There were women in nineteenth-century England without religion, but either they were few or they rarely felt obliged to publish an account of their lives. This in itself is a comment on the pervasive culture as it applied to women.

[44] The Revd. James Gardner, *Memoirs of Christian Females*, Edinburgh, 1841, p. 7.

[45] [Lewis], *Woman's Mission*, p. 11. See also, A Woman [Eliza Hinton], *Thoughts for the Heart. Addressed to Women*, London, 1847, pp. 28–9; *Howitt's Journal*, i, April 10, 1847, p. 29.

[46] I have included about 400 contemporary women's memoirs, autobiographies, biographies, etc. in the bibliography.

Remarks on the political life of the nation are uncommon in the autobiographical remains, particularly in the first half of the century. Waterloo, Peterloo, the New Poor Law, Chartism, elections, even reform bills seem not to have entered the consciousness of many of the female diarists and memoirists. Women were, of course, largely abstracted from the process of government and the springs of political action. Without the vote, usually ill-educated for discussions of political and social theory, and if married, with few property rights before 1882, they were badly placed to share in the excitement or the spoils of politics. Custom too isolated women from political reality, for men, especially 'gentlemen', avoided the serious discussion of issues in their company (artists and intellectuals were the exception). Thinking back on his Victorian upbringing, Bertrand Russell said: 'you had a whole different vocabulary when talking to women . . . you didn't dream of saying a word of truth to a woman. It would be gross, impertinent, and horrible. You would just lie blankly, and it was essential to good manners that you should.'[47] Given this state of affairs, it should not come as a surprise that the autobiographies of women, like women's literature generally, dealt largely with the affections, the duties of wives and mothers, and the conventional pattern of women's lives. The fact that there were important exceptions should not blind us to the truth of this generalization.

One thing revealed by the literature of and about women is that they were more likely than men to identify themselves with biblical characters. This, it was argued, flowed from their dependent nature, from their tendency to look up and see themselves in others.[48] Rarely enjoying an education that led to theology, women sought in Scripture what could explain or give meaning to their more mundane existence. To them religion tended to be personal and social, something to be integrated into daily life and acted upon. A Christian need only submit to Jesus and obey his moral laws; to refuse was simply impractical. The message of Christ was clear: the salvation of the world through service and self-sacrifice.

[47] 'Speaking Personally, Bertrand Russell', ed. John Chandos, Riverside RLP 7014/7015, side 3.

[48] William Landels, *Woman: her Position and Power*, London [1870], pp. 89–90.

After Christ, work turns to privilege;
And henceforth one with our humanity.[49]

A woman's emulation of Christ, remarked one theologian, joined with her 'instinctive tendency' in setting her down the philanthropic path. This, he added, was why women were more often benevolent than men.[50]

In their reading of the New Testament women sought and discovered a Christ who was sympathetic to their condition. (Looking for different things in Scripture, they did not always find what men found.) The Christ they revered was the atoning saviour of the innocent and craving heart, who treated women with the utmost respect and tenderness. He gave and demanded love and represented a God who was wholly love, who 'so loved the world that He gave his only begotten Son'. To women, Christ was, above all, a martyr to love. If there was a conviction peculiar to nineteenth-century philanthropic women it was their belief, inspired by Christ, that love could transform society. They did not always live up to this conviction, but their charitable work was 'undertaken most often from pure desire to help some one else to know something of the mysterious happiness of love'.[51] In two lines from her didactic poem 'The Rights of Woman' Mrs Isabel Reaney, propagandist for female philanthropy, summed up this view, written with the life of Christ in mind:

The Right to live for those we love;
The Right to die that love to prove.[52]

Their imitation of Christ, yet more immediately their identification with him, was the most crucial force behind the charitable activity of many women.

But Christ, 'the heavenly bridegroom', was not the only guide to conduct in the Bible; the women of Scripture could also shed light on woman's nature and mission. (Queen

[49] Elizabeth Barrett Browning, *Aurora Leigh*, London, 1857, p. 349.

[50] Landels, *Woman: Her Position and Power*, pp. 89–90; George Everard, *Bright and Fair. A Book for young ladies*, London, 1882, p. 23.

[51] *The Nineteenth Century*, vii, 1880, p. 620.

[52] Mrs G. S. Reaney, *Our Daughters: their lives here and hereafter*, London, 1881, p. xvi. 'The whole law of woman's life is a law of love', wrote Mrs Ellis, *The Daughters of England*, p. 23.

Victoria was perhaps the only woman who meant more as a model of conduct.) Nineteenth-century writers ransacked the Testaments for insights into female character and used biblical women to illustrate important principles of sacred authority.[53] Mary Magdalen at Calvary was the model of fidelity; Phoebe of Cenchreae, 'a servant of the church', compassion incarnate; Dorcas of Joppa, who made clothes for the poor, synonymous with good works; Rebekah was the personification of industry and piety; Lydia an example of benevolence and self-sacrifice; Priscilla an active Christian; Mary a contemplative one; Esther a patriot; and Ruth a friend. Biblical women, whose names were so commonly given to nineteenth-century children, became powerful images in the female mind, in the minds of everyone in the Christian world. Is not a beautiful woman with a child on her breast the symbol of charity in art?

The women of Scripture gave strength to the claims of their nineteenth-century disciples for a wider role in society, for they were sacred testimony that a 'virtuous woman' not only 'looketh well to the ways of her household', (Proverbs 31: 27) but furrowed in the Lord's vineyard beside Christ and his apostles. Phoebe, called 'deaconess' in the original Greek Version (Romans 16: 1), gave the deaconess movement in the Church of England its justification in Scripture, so crucial to the cause.[54] Other women had ministered to

[53] See, for example, Clara Lucas Balfour, *The Women of Scripture*, London, 1847; William Landels, *Lessons for Maidens, Wives, and Mothers, from some of the Representative Women of Scripture*, London 1865; John Angell James, *Female Piety; or, the Young Woman's Friend and Guide through Life to Immortality*, London, 1852; Mrs Elizabeth Sandford, *Woman, in her social and domestic character*, London 1831, p. 51; *The Quarterly Review*, vol. cviii, 1860, pp. 357–8; [E. Bickersteth], *Woman's Service on the Lord's Day*, London, 1861; The Revd. Isaac Williams, *Female Characters of Holy Scripture; in a series of sermons*, London, 1859; [Ellen Ranyard], *Life Work; or the Link and the Rivet*, London, 1861.

[54] Phoebe was invariably cited when justification for the deaconess movement was sought. See, for example, *The Quarterly Review*, cviii, 1860, p. 357; J. S. Howson, *Deaconesses; or the official help of women in parochial work and in charitable institutions*, London, 1862, p. 53; Mrs Susannah Meredith, *Wanted, Deaconesses for the Service of the Church*, London [1872], p. 6; T. T. Carter, *Is it well to institute Sisterhoods in the Church of England, for the care of female penitents?* London, 1851, p. 16; T. T. Carter, *Objections to Sisterhoods considered, in a Letter to a Parent*, London, 1853, p. 20.

Christ himself in Galilee. They had been last at the Cross and first at the sepulchre. In conforming to the active benevolence of Mary Magdalene and Mary the mother of James, Phoebe and Rebekah, Dorcas and Priscilla, women found their highest calling and assuaged the sin of Eve.

So Christianity confirmed what nature decreed: women had a rightful and important place in the charitable world. But to most authorities, particularly the men, it was a subordinate place. To them, God ordained that woman in her public relations, as in her domestic ones, would be the helpmeet of man; the 'gifts and graces' peculiar to her were to be cultivated 'simply as a means of helping'.[55] The running of a philanthropic society could be compared to the running of a family: men were to provide the intelligence and direction, women 'the better heart, the truer intuition of the right',[56] and not least the unflagging industry that kept the institution together. So the gentle and compassionate sex, given to benevolence by nature, from childhood taught to bear the burden of good works, discovered, not to their surprise, that their activities in the powerful world of organized charity were to be hedged in by restrictions; their contribution might be superior in degree but it would be inferior in kind. There was ample room for self-styled Phoebes, Dorcases, and Rebekahs in the nineteenth-century vineyard, but they must not forget the dependence and humility becoming a female labourer in the service of their master. This was a view that women came to challenge.

[55] Mary Anne Schimmelpenninck, *Sacred Musings on Manifestations of God to the Soul of Man.* London, 1860, p. 291. Sandford, *Woman, in her social and domestic character*, p. 154.

[56] Theodore Parker, *A Sermon on the public function of Woman*, Boston, 1853, p. 19.

PART ONE:

The Power of the Purse

I

The Advance of Women as Contributors

Social historians may sometimes share Walter Bagehot's melancholy doubt 'whether the benevolence of mankind does most harm or good', but they do not dispute that philanthropists were ubiquitous in nineteenth-century England.[1] As befitted a nation which believed that philanthropy was the most reliable and wholesome remedy for its ills, there was a phenomenal growth of charitable funds and institutions.[2] Individuals contributed staggering sums: the patent medicine vendor Thomas Holloway well over a million pounds, Angela Burdett-Coutts, whose profession was philanthropy, perhaps more.[3] Down the social scale, it was not uncommon for families to tithe their income to charitable causes. A survey of forty-two middle-class families in the 1890s showed that they spent more on charity than on rent, clothing, servants' wages or any other item except food.[4]

The amount of money contributed each year to charity, not including donations at the altar and unremembered alms, far exceeded the gross expenditure on poor relief.[5] *The Times* reported in 1885 that charitable receipts in London alone came to more than the national budgets of Denmark, Portugal, Sweden, or the Swiss Confederation.[6] There were nearly 1,000 charitable institutions in London by the end of the century with an income well in excess of £6,000,000.[7]

[1] Walter Bagehot, *Physics and Politics*, London, 1872, pp. 188–9.

[2] See David Owen, *English Philanthropy, 1660–1960*, Cambridge, Mass. and London, 1964; Ford K. Brown, *Fathers of the Victorians*, Cambridge, 1961.

[3] Owen, *English Philanthropy*, pp. 401, 413–20.

[4] On average these families expended 10.7% on charity. See [E.Grubb?] *Statistics of Middle-Class Expenditure*, British Library of Political and Economic Science, Pamphlet HD6/D267, Table IX.

[5] Geoffrey Best, *Mid-Victorian Britain, 1851–1875*, London, 1971, p. 140.

[6] Jan. 9, 1885, cited in Owen, *English Philanthropy*, p. 469.

[7] W. F. Howe, *Twenty-Fourth Annual Edition of the Classified Directory to the Metropolitan Charities for 1899*, London, 1899, p. xxvi. Howe listed 970

This is not to say that the English were more benevolent in the nineteenth century than ever before or that philanthropic contributions rose at a higher rate than the national income. This may well have been the case, but it would be difficult to prove from the charitable statistics available for the nation at large in the years before 1900.[8]

If we know little about the level of nineteenth-century benevolence compared to earlier periods, we are on firmer ground when discussing the structure of philanthropy. Voluntary societies, financed by individual contributions and governed by committees, existed before 1800, but the nineteenth century saw them come 'to full, indeed almost rankly luxuriant, bloom'.[9] One contemporary called the period the age of charitable societies: 'for the cure of every sorrow . . . there are patrons, vice-presidents, and secretaries. For the diffusion of every blessing . . . there is a committee.'[10] Centralized and often highly specialized, the expanding charitable institutions combated a host of human ills, moral and physical, individual and social, many of them associated with the increasingly urban and industrial environment. Casual almsgiving and the visits of vicars and lady bountifuls, which helped to ameliorate poverty and distress in pre-industrial England did not die out, but they were inadequate to deal with the changing conditions. In the rapidly growing cities, for example, knowledge of hardship was not so easy to come by, and it was more difficult to distinguish between real and feigned distress. The voluntary societies sought to remedy these and other problems. They intervened in the relationship between the benevolent and the needy, laid down and administered policy, encouraged new methods of fund-raising, and in many cases created a system of auxiliaries which extended their influence through the country.

Many of the issues raised by the flowering of nineteenth-

metropolitan charities and a figure of £6,207,291, but this amount came from only 736 of the charities; the remaining 234 did not provide him with financial information.

[8] Owen, *English Philanthropy*, p. 470. See also Brian Harrison, 'Philanthropy and the Victorians', *Victorian Studies*, ix, June 1966, p. 373.

[9] Owen, *English Philanthropy*, p. 92.

[10] Sir James Stephen, *Essays in Ecclesiastical Biography*, 2 vols., London, 1849, i, p. 382.

century charities have been treated with considerable insight. But the role of women in these institutions has not attracted very much attention. Little is known about the origins or scope of their participation; and no statistical study has been carried out which would provide information on the numbers of women contributing and the amounts they donated to the various charities.[11] The distinguished preacher Rowland Hill said at the inaugural meeting of the Southwark Bible Society in 1812 that 'the ladies do not generally carry the larger purse' and that their contributions to philanthropy were little more than 'pin-money' vouchsafed them by their good husbands.[12] Is his opinion borne out by the evidence and does it hold true for the nineteenth century generally? Only a study of charitable statistics can tell us.

It is best to preface the statistical material with a discussion of the activities of women in the auxiliary movement, for it was often at the local level that they took their first, frequently timid steps in charitable institutions. The auxiliary system proved highly successful in marshalling the support of women, despite some initial difficulties. No other charitable innovation brought in such massive sums of money, commonly over half of the income of those societies which adopted the system. Pioneered by the missionary and Bible societies, the idea can be traced at least as far back as 1792, when the French Bible Society of London recommended 'the formation of societies in different parts of the country, to assist them in the attainment of their object'.[13] But 1792 was not a very propitious year to promote Christianity in France and so the idea remained on that society's drawing-boards. Just when and where it was first put into practice is difficult to

[11] Working primarily from wills, W. K. Jordan has put together some interesting statistical material on the role of women in philanthropy in the years 1480–1660. See *Philanthropy in England, 1480–1660*, London, 1959. See also F. K. Prochaska, 'Women in English Philanthropy, 1790–1830', *International Review of Social History*, xix, 1974, pp. 426–45.

[12] *Report of the Proceedings at the Horns Tavern, Kennington, . . . when an Auxiliary Bible Society was established, for Southwark and its Vicinity*, Southwark, 1812, p. 31.

[13] Quoted in C. S. Dudley, *An Analysis of the system of the Bible Society throughout its various parts*, London, 1821, p. 135.

determine, but it is likely that auxiliaries established by men came into being before those established by women.

Perhaps the first auxiliary established by women was the 'Female Missionary Society' in Northampton, which contributed 10*s*.6*d*. to the Baptist Missionary Society in 1805.[14] Another early example was the working-class Female Servants' Society, which opened in Aberdeen in 1809, to assist the Edinburgh Bible Society.[15] At about the same time penny societies of ladies and children formed in the London Society for Promoting Christianity amongst the Jews.[16] The first women's association in 'direct and exclusive' connection with the British and Foreign Bible Society, which probably did more than any other institution to spread the auxiliary idea, opened in Westminster in 1811.[17] If these were not the earliest of the female auxiliaries in Britain, they were among the earliest. By the 1840s, female branches, sometimes called auxiliaries, sometimes associations, were part of most of the charities which sought nation-wide support, whether they championed the cause of chimney-sweeps or African slaves. They were tendrils of the charitable societies, shooting across the country and depending on local benevolence for sustenance. Some of them consisted of only a few women; others recruited hundreds of members. Run by committees made up of secretaries, treasurers, and members who subscribed a certain sum, they conformed to rules coming out of headquarters; their essential purpose was 'to solicit and obtain patronage'.[18]

How did these female auxiliaries get their start? Often a woman sympathetic to the cause of a particular society, perhaps on hearing a sermon or after reading a pamphlet, rallied her family and friends. She could then write directly to the charity or contact her local minister who would write

[14] *Periodical Accounts relative to the Baptist Missionary Society*, iii, London, 1806, p. 138.

[15] Dudley, *An Analysis of the system of the Bible Society*, pp. 355–6.

[16] *The Third Report of the Committee of the London Society for Promoting Christianity amongst the Jews*, London [1811], pp. 180–2.

[17] Dudley, *An Analysis of the system of the Bible Society*, p. 357.

[18] Ibid., p. 375. For another set of rules see *The Annual Report of the Baptist Missionary Society for the Year ending March the Thirty-First, M.DCCC.L.*, London, 1850, p. 88.

on her behalf. One method was for the woman organizer to find ten women, who were to find ten others, each of whom would enlist a fixed number of subscribers.[19] As time passed, more and more auxiliaries resulted from programmes initiated by the charities themselves. Most of the larger societies had special agents who travelled the country promoting female participation. Perhaps the most notable of these people in the early years of the century was the Quaker, Charles Dudley, an avid supporter of women's work in the British and Foreign Bible Society. He met and corresponded with women around Britain and often was called in by them to revitalize a declining association or to found a new one. In this work he was ably assisted by women like Miss Maria Hope of Liverpool, who organized ladies' branches for the Bible Society in the north. Miss Hope and Charles Dudley, and many like them in other charities, spread the message, encouraged competition for funds between different female associations, and counted the money.[20]

There was considerable opposition to women's auxiliaries, and much of this came from conservative churchmen. They warned of vain and unfeminine women, 'feverish' for publicity, who neglected their domestic duties, of women too stupid to administer efficiently or even add up their collections correctly, of 'amazonian women who challenge attention, and put us upon our defence'.[21] Nor were women spared the accusation of fanaticism. According to one critic a group of ladies forced their way into a clergyman's kitchen and alienated his entire household by their

[19] British Library, Add. Mss. 18,591, ff. 117–19; Joyce Hemlow, *The History of Fanny Burney*, Oxford, 1958, pp. 243–4.

[20] The Home Correspondence Inwards of the British and Foreign Bible Society has a number of letters dealing with women's auxiliaries.

[21] Revd. Richard Lloyd, *Strictures on a recent publication entitled 'The Church her own Enemy'; to which are added, a refutation of the arguments contained in the Rev. Edward Cooper's Letter to the Author; and an admonitory address to the female sex*, London, 1819, p. 113; H. H. Norris, *The Origin, Progress, and Existing Circumstances, of the London Society for Promoting Christianity amongst the Jews*, London, 1825, p. 239. The British and Foreign Bible Society Archives contains several letters from churchmen complaining of women's auxiliaries. See, for example, Home Correspondence Inwards, The Revd. R. B. Fisher to J. Hughes, 18 March, 1817, and Liverpool Auxiliary Committee to BFBS Committee, 14 October, 1819. See also *The Anti-Jacobin Review*, lviii, 1820, p. 63.

tub-thumping.[22] 'As soon as the herald of the morn has aroused the associated ladies from their pillows', complained another churchman, 'they issue from their homes, equipped, like so many Excise-officers, with account-books in their hands, and lightly trip from street to street, and from door to door, in the glorious office of collecting subscriptions from the poor and indigent'.[23] Was not a man's home his castle, to be defended from such 'presumptuous' females, who acted as a kind of 'theological police'?[24] At their most extreme the critics saw the 'associated ladies' as a threat to family life, ecclesiastical efficiency, and an ordered society.

The women stood their ground. Some of them, we are told, even threatened to leave their husbands rather than leave their associations.[25] More commonly they compared the fruits of their philanthropic labour with those of the associated men, who were not widely known for their effectiveness. 'Defects in the plan of doing good can be of no moment compared with a plan which does no good at all' remarked one defender of women's associations.[26] Most of the men's societies that used auxiliaries made judicious compromises, which were meant to satisfy masculine sensibilities without alienating feminine support. As Jabez Bunting told the ladies at a meeting of the Methodist Missionary Society in Leeds, women subscribers were preferable to women speakers.[27] To please those who valued retirement in the ladies, some societies excluded them from public meetings. Women could not attend the annual general meeting of the Church Missionary Society until 1813.[28] In the Bible Society they had to wait until

[22] Bodleian Library, W. Conybeare to Mrs Hodge, n.d., Montagu Mss. d. 12, II, f. 131, cited in R. H. Martin, 'The Pan-Evangelical Impulse in Britain 1795–1830: with special reference to four London Societies', Oxford D. Phil. Thesis, 1974, p. 224.

[23] A Churchman, *A Letter to the Church Members of the Auxiliary Bible Society, Liverpool*, Liverpool, 1819, p. 7.

[24] Ibid., pp. 9–10. [25] Ibid., pp. 18–19.

[26] BFBS Archives, Home Correspondence Inwards, Ms Letter, Liverpool Bible Auxiliary to BFBS parent committee, 14 Oct. 1819, f. 4.

[27] James Nichols, *A Report of the Principal Speeches delivered on the Sixth Day of October 1813, at the Formation of the Methodist Missionary Society, for the Leeds District*, Leeds, 1813, p. 47.

[28] Eugene Stock, *The History of the Church Missionary Society*, 4 vols., London, 1899–1916, iii, p. 820.

1831.[29] And in the Society for the Propagation of the Gospel they were admitted in the 1820s but had to be discreetly hidden from view.[30] To safeguard themselves further from 'ostentatious' and 'amazonian' women, committee-men often grafted female associations on to men's auxiliaries, thereby guaranteeing that women were kept under some control. Another protection thought necessary was to have male agents visit the provinces to ensure that women carried out the rules of the particular society to the letter.

As these agents discovered, women were rather good at their work. Enthusiastic and with time on their hands, they commonly took over men's auxiliaries and made them profitable. Moreover, as Charles Dudley argued, women were not so prone as men to abuse their office.[31] He concluded that associations 'can be best—indeed I may say *only* conducted by Females'.[32] And he added: 'If any human means have contributed to set the seal of *permanence* on the Bible Society, it is the application of *Female Agency*.'[33] In 1819, two years after he made these remarks and only eight years after the first women's association joined the Bible Society, the committee of that great charity declared that while they regretted the 'evils and inconveniences' of the ladies' associations, they felt 'so grateful for the immense advantages already derived from that source' that they could not refrain from recommending their continuation.[34] This pronouncement was not unconnected with the fact that by 1819 there were 350 female associations in the Bible Society with about 10,000 women 'regularly employed'; they were bringing in tens of thousands of pounds each year.[35] This was quite a sum from 'pin-money'. The critics retreated.

[29] Brown, *Fathers of the Victorians*, p. 249.

[30] *The Christian Observer*, Jan. 1861, p. 40.

[31] BFBS Archives, Home Correspondence Inwards, Charles Dudley to Anon., 14 May, 1817.

[32] Ibid., Charles Dudley to J. Tarn, 4 October, 1817. See also, *Bible Associations Exposed. Being a Review of the Fourth Annual Report, of the Committee of the Henley Bible Society*, London, 1818, p. 7.

[33] BFBS Archives, Homes Correspondence Inwards, Charles Dudley to Anon., 14 May, 1817.

[34] Ibid., Ms Minutes of Committee, Nov. 15, 1819, f. 204.

[35] Lloyd, *Strictures on a recent publication*, pp. 113–14, gives a figure of 14,000 women workers in the association movement in the Bible Society by 1820. Dudley, *An Analysis of the system of the Bible Society*, p. 501, gives a figure of 10,000 in 1821. See also the Annual Reports of the Bible Society.

Such was the power of the purse in the Bible Society; and as a look through the charitable reports of other institutions with a similar administration will show, it was not alone in receiving great benefits from its women's branches, nor alone in appreciating them. The London Society for Promoting Christianity amongst the Jews recognized by 1825 that a 'very principal portion' of its income came from female auxiliaries.[36] Nor did the historian of the Church Missionary Society exaggerate when he wrote that in the early years his institution 'largely depended' on women's associations for the collection of its funds.[37] 'The managers of nearly all our societies', wrote another authority, 'find in women their chief auxiliaries; and in their straits it is to woman's help that they mainly look for being brought successfully through'.[38] These remarks should be kept in mind as we turn to the subscription and legacy lists of the charitable societies.

For most of the thousands of charitable institutions in nineteenth-century England the records are scant; long runs of annual reports are the exception rather than the rule; and long runs which include detailed lists of subscribers are even more exceptional. None the less quite a few societies published lists of subscribers for two or more years; a study of these should enable us to make a fair assessment of the pattern of female contributions. Fifty such charities have been found and are included in appendix I. Their committees were composed entirely of men, though some of them received assistance from women, either as patronesses or in sub-committees. (More will be said later about charities with mixed committees of men and women and about those run exclusively by women.) In size and wealth these fifty better documented societies have a greater significance than their numbers suggest, for they include many of the largest institutions, whose funds accounted for a high percentage of the charitable receipts in any given year. Furthermore, they cover a wide range of

[36] Quoted from the fourteenth annual report of the society in Norris, *The Origin, Progress, and Existing Circumstances, of the London Society for Promoting Christianity amongst the Jews*, p. 239. The extent to which this society was indebted to women is suggested in the Ladies' Committee Minutes (1825–37), Church Mission to the Jews, Bodleian Library.

[37] Stock, *History of the Church Missionary Society*, iii, p. 321.

[38] Landels, *Woman: Her Position and Power*, p. 127.

philanthropic activity, both in type and in affiliation. Most of them had their headquarters in London; and while many of them had a system of provincial auxiliaries, the subscriptions which have been counted are normally limited to those sent directly to the parent society.[39]

All of the fifty societies listed in appendix I were growing during the nineteenth century, some more rapidly than others. But a more remarkable feature was the dramatic rise in the percentage of women contributors within these expanding charities. In forty-four out of the fifty the percentage of female subscribers increased, and in the majority of these the increase was considerable. This was particularly true of the largest charities, for example the missionary and Bible societies. In the Baptist Missionary Society, the British and Foreign Bible Society, the Church Missionary Society, and the London Missionary Society the percentages of female subscriptions rose respectively from 15 to 43% (1800–1900), 12 to 40% (1805–95), 12 to 49% (1801–1900), and 17 to 44% (1820–1900). In some of the other large societies the percentages rose even higher. At the end of the century women made up 57% of the subscribers to the London City Mission, a leading visiting charity; in the London Society for Promoting Christianity amongst the Jews the percentage reached 59%; and in the Royal Society for the Prevention of Cruelty to Animals the figure rose to 69%. In the six societies in which the percentages of female subscribers did not rise, it remained constant in two of them and dropped only slightly in three others. But it should be added that the level of female participation in these charities was relatively high in the earlier lists. In the Philanthropic Society, an institution for juvenile offenders, where the decline in female contributors was more significant, from 23 to 14% (1816–48), it is worth noting that this charity phased out the girls' side of its activities by 1845. It could be inferred from this that women subscribers did not give their money away indiscriminately.

[39] The contributions from provincial auxiliaries are particularly difficult to deal with, for not only do they take up page after page of fine print but they often include donations which cannot be attributed with certainty to one sex or the other, for example prayer meetings, Sunday schools, charity sermons, and collecting cards.

In addition to these fifty societies another sixty-five have been found with one subscription list (see appendix II). It is, of course, impossible to establish any trend in female contributions from these lists, but they do give an impression of the kind of benevolence which interested women. Here, as in the aforementioned charities, the managing committees were made up exclusively of men, though women served in many of them as patronesses and on sub-committees. Their role depended largely on the nature of the society and on the management's connections, and as might be expected the percentage of female subscribers varied considerably. In the miniscule Institution for Rendering Assistance to Shipwrecked Mariners, the sole female subscriber in 1809 was its patroness, the Princess of Wales. Perhaps because of their sensibilities, women showed little interest in foreigners and the ruptured poor: they made up only 4% of the subscribers to the Society of Friends of Foreigners in Distress in 1814 and only 3% of those to the National Truss Society in 1842. Women preferred to contribute to those charities which dealt with pregnancies, children, servants, and the problems of aging and distressed females. In the Aged Female Society in Sheffield, for example, they made up 79% of the subscribers in 1827, and in the Institution for the Employment of Needlewomen 75% in 1864. Vice, or its elimination, also caught their fancy, for as early as 1802 they made up 31% of the subscribers to William Wilberforce's Society for the Suppression of Vice. These figures, like the earlier ones listed, suggest that women did not give their money thoughtlessly. They invested in charity and expected a return that related to their experience and to their own sex. This was only natural in a society which made such distinctions between men and women.

Before moving on to the question of the contribution of women in purely financial terms, it is worth citing some subscription lists from charities in which men and women worked together on the same managing committee and from a few charities managed exclusively by women. In the first half of the century societies with a mixed executive of men and women were rare.[40] But in the mid- and late Victorian years, in what

[40] The earliest example which I have found of a charity with a mixed management of men and women was the Ladies' Charity, Liverpool, established

may be seen as a tribute to the rising voice of women in organized philanthropy, committee-men frequently welcomed female members to their ranks. This was particularly evident in moral reform societies and in charities dealing with the problems of women and children. But in the last years of the century even a few of the more traditional institutions, the Royal Society for the Prevention of Cruelty to Animals and the London Missionary Society for example, admitted women to their inner councils.[41] Not surprisingly, the percentage of female subscribers to charities which welcomed women to management tended to be higher than to those from which they were excluded (see appendix III). In the Workhouse Visiting Society, for example, they made up 84% of the subscribers in 1864, in the Society for Promoting the Employment of Women 80% in 1867, and in the Moral Reform Union 77% in 1883.

These figures are somewhat misleading, for the societies in which men and women worked together in committee were not invariably bastions of male supremacy to which women were invited. Many of them were female charities to which men were invited. Just as committee-men were becoming more and more likely to introduce women to their council-chambers, so too were committee-women introducing men to theirs. No one would describe such institutions as the Women's Protective and Provident League, the Society for Promoting the Employment of Women, or the Ladies' National Association for the Repeal of the Contagious Diseases Acts, for example, as men's societies; yet they had men on their committees. Such organizations were in a long and distinguished tradition of female charities dating back to the eighteenth century. In the late Victorian years, institutions with a mixed management were still not so numerous

in 1796. See *The Eleventh Report of the Ladies' Charity; or Institution for the Relief of Poor Married Women in Child-bed, Liverpool*, Liverpool, 1807.

[41] Women joined the committee of the RSPCA in 1896. After many years of service in the London Missionary Society, women attended their first meeting as directors on May 26, 1891. See the LMS Board Minutes, Box 45, May 26, 1891, School of Oriental and African Studies Library. The LMS was the only one of the great missionary societies to permit women to join its managing committee in the nineteenth century. In the British and Foreign Bible Society the first woman joined the committee only in 1944.

as those managed by one or the other sex. But they reflected a growing improvement in the business relations between men and women and a willingness on the part of both sexes to work together in a benevolent cause. Emily Davies wrote to Barbara Leigh Smith (Bodichon), philanthropist and suffragist, in the 1860s: 'There is nothing at all new in women's working together. All over the Country, there are Ladies' Associations, Ladies' Committees, Schools managed by ladies, Magazines conducted by ladies etc. etc., which get on well enough. The new and difficult thing is, for men and women to work together on equal terms.'[42]

What can only be described as an explosion of charities managed exclusively by women took place in the nineteenth century. They begin to appear in increasing numbers in the early part of the century; by the mid-Victorian years there were hundreds of them, not including auxiliaries.[43] They covered a very wide range of interests, but most notably causes relating to women and children. Much more will be said about some of them later, but for now it is worth noting the percentage of female subscribers in a few of them (see appendix IV). The Ladies' Benevolent Society, Liverpool, the Friendly Female Society, and the Women's Vegetarian Union were typical in having over 80% of their subscriptions from women. This sample confirms the not very startling conclusion suggested by the previous samples, that the more power women held in a society the more likely they were to contribute to it. The emergence of female charities, moreover, must have drained off some of the support that women might have been expected to give to the men's societies. This makes their contribution to the charities managed by men all the more impressive.

What about the contribution of women to charitable institutions in purely financial terms? This is rather more difficult to assess, for some societies did not include the amount of contribution in their subscription lists; in many

[42] Emily Davies to Barbara Leigh Smith Bodichon, Jan. 14 [1863?] Bodichon Papers, Girton College, Cambridge. I would like to thank Diane Worzala for this reference.

[43] There is more information on the early female societies in Prochaska, 'Women in English Philanthropy, 1790–1830', *IRSH*, pp. 430–1.

others the sheer bulk of information makes the task of isolating the figures for women a formidable one. It may be helpful, however, to look at twenty charities, all managed exclusively by men, in which the evidence is relatively easy to manage (see appendix V). In eleven of these societies, including the Home Missionary Society and the Church Lads' Brigade, the financial contribution of women was lower than their numerical percentage. In the nine others, however, including the Irish Evangelical Society and the Aborigines' Protection Society, women paid their share and more. Such a sample is far from exhaustive, but it is fair to conclude from it that the financial contribution of women to institutional charity was very nearly in proportion to the number of women members.

We must not overlook the importance of legacies in assessing the financial contribution of women to philanthropy. They are not, as David Owen discovered, an 'infallible yardstick for measuring the dimensions' of nineteenth-century charity,[44] but they can be a very useful tool in quantifying the female contribution to individual charities. Twenty-nine societies have been found which published information on legacies (see appendix VI). Most of these charities have been examined already for their female subscribers (see appendices I and II), and thus we can compare the percentages of female legacies with subscriptions. As with the female subscriptions the number of legacies from women rose during the century in virtually all of the societies studied; this is obvious from a cursory look at the long rolls of legacies published. But more interesting is the relatively high percentage of female legacies compared to the percentage of female subscriptions. In the Baptist Missionary Society, for example, 56% of the legacies were from females (79% in financial terms) in the years 1895–1900; female subscriptions made up just 43% of the total in 1900. Women made up 66% of the legators in the London City Mission between 1865 and 1870 (69% in financial terms), while in 1870 female subscribers made up 42%. And in the Lord's-Day Observance Society between 1844 and 1876, 59% of the legacies were

[44] Owen, *English Philanthropy*, p. 471.

from women (58% in financial terms); only 33% of the subscribers were women in 1876. Most of the other charities examined reveal a similar pattern; the percentage of female subscriptions rarely matched the percentage of female legacies. This disparity suggests that the percentage of female subscribers may have been artificially low; if so, it was probably the result of putting female contributions in the husband's name.

Women contributed vast sums to nineteenth-century societies in the form of bequests, but did they leave a larger portion of their estates to charity than men? This question can only be answered by a study of wills. In the first sample one hundred wills from the 1840s have been examined, fifty from men and fifty from women.[45] The estates of the men were larger on average than those of the women, as one might expect, but the percentage of the assets given to charity was a different matter:

	Fifty Men	Fifty Women
Value of combined estates:	£1,162,650	£595,850
Left to charity:	£52,510	£68,850
Percentage left to charity:	4.5%	11.6%
	Twenty-Nine Spinsters	Twenty-One Widows
Value of combined estates:	£255,550	£340,300
Left to charity:	£38,461	£30,389
Percentage left to charity:	15.1%	8.9%

Not only did the women leave much more to charity than the men on average, but unmarried women left rather more than widows.

This view is reinforced by a sample of wills from the 1860s,

[45] These wills are all from the Records of the Prerogative Court of Canterbury in the Public Record Office (Prob 11). They have been examined in association with the relevant Probate Act books, which provide estimates for the gross unsettled personalty of the individual estates (Prob 8). The names were taken from the legacy lists of several London charities, a method which guarantees that the bequest was made and not simply intended at the time of the drawing-up of the will.

published in the *Illustrated London News*.[46] Again one hundred wills containing bequests to charity were examined, fifty from men and fifty from women. These estates tended to be larger than those studied from the 1840s; their size, of course, was a reason why the *Illustrated London News* thought them worth reporting:

	Fifty Men	Fifty Women
Value of combined estates:	£6,263,000	£2,253,000
Left to charity:	£210,660	£292,335
Percentage left to charity:	3.4%	13%

	Twenty-Nine Spinsters	Twenty-One Widows
Value of combined estates:	£1,191,000	£1,062,000
Left to charity:	£181,025	£111,310
Percentage left to charity:	15.2%	10.5%

Finally, figures from the *Daily Telegraph* confirm the view that women were much more likely than men to leave a larger portion of their assets to philanthropy. This newspaper listed fifty or sixty wills annually in the 1890s, the total number coming to 466.[47] These combined estates came to nearly £76,000,000, of which £10,200,000, or over 13%, went to charity. In a study of these figures, *The Times* pointed out that the 150 women in the sample willed, on average, 25.8% of their estates to benevolent causes, the 316 men only 11.3%.[48] We should not infer from these figures that men and women were more benevolent in the 1890s than they were in the 1840s and 1860s, for the *Daily Telegraph* selected these wills for inclusion because of the large bequests made to charity. Commenting on the results of this survey, *The Times* remarked that only occasionally did the English leave funds to charity if they had families to provide for, and that for the most part such bequests came from persons who had been philanthropic in life. David Owen, who

[46] These wills were taken as they appeared, for the years 1862–8.

[47] The *Daily Telegraph* listed these wills in the last week of December, 1891 to 1898.

[48] *The Times*, 25 September, 1899.

cites the figures from the *Daily Telegraph*, added that women were more generous 'because of the large number of widows and spinsters without family obligations'.[49] This is probably true, and it helps to explain why spinsters were rather more generous than widows in the samples from the 1840s and 1860s.

The statistical evidence, which is supported by the ephemeral literature, suggests that the contribution of women to institutional charity began to rise markedly in the early nineteenth century. (In some societies this trend can be detected in the late eighteenth century.) It has been argued elsewhere that the French Revolution and the Napoleonic wars enhanced the status of women and stimulated their interest in good works, and undoubtedly many an English woman's subscription resulted from the war with France and the consequential suffering.[50] There is no doubt some truth in the old cliché that the English became pious when the French became republican; such piety worked not only to release reserves of the English woman's wealth, but also enhanced the conservative colour of her philanthropy. Years later the Crimean War also stimulated female charitable contributions, particularly to hospitals and to nursing institutions. In many societies women appeared on the scene quite dramatically, for example in those which sprang up during an epidemic or a hard winter. The cholera outbreak of 1866, for instance, resulted in the creation of several women's charities.[51] One could also cite societies in which women, perhaps too amateurish and too sentimental, lost interest rather quickly. Such occurrences led John Stuart Mill, among others, to complain of the unreliability and 'evil tendency' of female philanthropy.[52] But despite such cases, the financial records of the largest and most long-lived societies reveal a steady advance in women's contributions after the pronounced rise

[49] Owen, *English Philanthropy*, p. 471.

[50] Prochaska, 'Women in English Philanthropy, 1790–1830', pp. 438–40. The English civil war had something of the same effect, see Jordan, *Philanthropy in England, 1480–1660*, p. 355, note; and also W. K. Jordan, *The Charities of London, 1480–1660*, London, 1960, p. 29.

[51] See pages 160–1.

[52] John Stuart Mill, *The Subjection of Women*, London, 1869, p. 163.

in the early years of the century. Nor is there any reason to believe that female enthusiasm waned before 1900. There were more permanent forces at work pushing women in the direction of institutional charity than Napoleon and epidemics.

Within this framework of a steady increase in female contributions to philanthropy generally, women did shift their charitable priorities as time passed. The perennial human problems, disease, lying-in, and old age, and the perennial vices, drink and immorality, held their attention at high though varying levels right through the century. But some areas of concern diminished in importance, anti-slavery and the cause of climbing-boys for example; other problems became more prominent. As we shall see, women turned their attention to prison-visiting in the 1810s and 1820s, only to find their influence decline after the government inspectors moved in in the 1830s. In the 1850s and 1860s workhouse-visiting became something of a rage, not unconnected with the effects of the New Poor Law and the work of the National Association for the Promotion of Social Science, in which women played an active part. In the same years rescue work and the setting up of Magdalene homes and reformatories caught the imagination of benevolent women. This, in turn, stimulated their campaigns for moral and legal reform in the 1870s and 1880s.

For sheer variety the mid-Victorian period was perhaps the heyday of female charitable activity. Many of the societies founded in these years reflected the lively discussion of female rights and duties going on at the time; the Society for the Employment of Women and the National Society for Women's Suffrage spring to mind. It should be said again, however, that there was no apparent decline in women's overall contribution to charity during the nineteenth century. Philanthropic priorities varied from time to time, as one would expect in a nation undergoing such fundamental social and economic change, but the financial contribution of women to charity as a whole continued to increase. There appeared to be no incompatibility between their growing public role in local government and the paid professions toward the end of the century and the philanthropic role in which by this time they were established.

While an unprecedented and ever-increasing number of women contributed to the charitable societies in the nineteenth century, we must be cautious in making too much of this. Just as we know little about the level of nineteenth-century philanthropy generally compared to earlier periods, we know little about the level of female benevolence compared to previous centuries. For a start, we have very little information on the amount contributed by eighteenth-century women to charity, even less on the share of their total wealth that it constituted. Moreover, we would not always be comparing like with like, for the voluntary society did not typify eighteenth-century philanthropy to anything like the same extent as it did that of the nineteenth century. Before the rapid growth of voluntary societies, women more often may have used the traditional means of redistributing wealth, church and chapel collections or the casual distribution of alms, a practice which came under attack in the nineteenth century. It is tempting to infer from the subscription and legacy lists that Victorian women were more philanthropic than their grandmothers; yet the difference between them may have been largely a matter of the way in which their money was extracted or distributed, not one of generosity.

Having said that, it must be argued that the peculiar configuration of economic, social, and religious change taking place in nineteenth-century England appears to have heightened the philanthropic impulses of women. Certainly the charitable organizations were quick to turn these impulses to their advantage. The evangelicals, for our purposes all Bible Christians, were particularly adept at extracting money from women. They were not only relatively open to women socially, but they also worked harder than anyone else. This combination of zeal and organizational flexibility triggered the rapid expansion of female contributions in their charities. A look through the subscription and legacy lists shows that the rise in women's donations was considerable in those charities which justly can be called evangelical, most of the missionary and tract societies for example. It is interesting to note that women gave a larger proportion of the funds of the Religious Tract Society than of the

Society for Promoting Christian Knowledge and that in the National Society for Promoting the Education of the Poor, another society not usually associated with the evangelicals, the level of female participation was relatively low.

The charitable reports and ephemeral literature suggest that different denominations had different philanthropic emphases. Evangelical Quakers were especially prominent in the Peace movement; Unitarians often supported educational charities; and Low-Church Anglicans enthusiastically promoted tract and Bible societies. But no one denomination had a monopoly over any single charitable enterprise. Moreover, some sects were themselves divided over the issue of good works. The evangelical wing of the Quakers, it has been argued, had more in common with other non-Quaker evangelicals than with the Quietists in their own ranks, who were rarely active in philanthropic circles.[53] In many charities, some of which have been too readily labelled evangelical, middle-of-the-road and High-Church Anglicans joined in the common cause. While evangelicals set the pace, many Middle- and High-Church women, as well as Unitarian, Catholic, and Jewish women, were willing contributors to a great variety of charitable institutions. While sectarian specialities can be detected, there is little to suggest that women of one particular denomination were conspicuously more active in philanthropy generally, at least among the evangelicals.

Virtually all women, evangelical or atheist, rich or poor, felt the pressure to contribute. It was unrelenting. It came from the pulpit and the platform, the reports and pamphlets of the charitable societies, the numerous family and women's magazines, and from millions of penny tracts pumped out by the religious publishing houses. Crusading women writers like Hannah More early in the century, Anna Jameson in the 1850s, and Ellice Hopkins later on, were highly influential propagandists for the cause. Agents of the charitable societies put the public under the most direct pressure, especially the women and children who systematically canvassed their districts for donations. And men like Sir Thomas Bernard of the Society for Bettering the Condition of the Poor and

[53] Elizabeth Isichei, *Victorian Quakers*, London, 1970, p. 214.

James Mackintosh of the West London Lancastrian Association, not to mention many churchmen, added their weight by scouring the nation's drawing-rooms in the hope of persuading rich women to contribute to their respective causes.[54] They discovered what Edward Wakefield recognized in 1813 when he wrote to Francis Place about the income of the West London Lancastrian Association: 'ladies will be the most zealous instruments of which to avail ourselves'.[55] The man who could attract the ladies and tap their wealth was a common feature of the charities managed by men, and in order to open purses he sometimes had to make charity sound like an amusement, 'the gaming table without its horrors' to use Bernard's phrase.[56] All in all, female benevolence had never before been solicited on such a scale.

Whatever form it took the campaign to lighten the female purse was a phenomenal success. Of course, many of those who joined the bandwagon simply turned privilege into virtue and made the subscription resemble an indulgence. Indeed, it can be argued that charity turned into a branch of fashion in the nineteenth century. The long lists of subscribers, which commonly took up the bulk of the annual charitable reports, were in themselves a sign of status seeking. 'I find there is a good deal of sense in what you were saying yesterday as to giving money away', remarked one character from fiction. 'I have been looking at these applications and I see that the best people about here give very regularly.'[57] Yet it would be unfair to malign the many women, some privileged, others not, who contributed to charity from more noble motives. These were the women who, according to the impoverished poetess Mrs Maddocks, found their chief delight in 'blessing others'.[58] Sarah Trimmer and Hannah

[54] Thos. Frognall Dibdin, *Reminiscences of a Literary Life*, 2 Parts, London, 1836, i, p. 231; Edward Wakefield to Francis Place, 1 December, 1813, British Library, Add. Mss, 35,753, ff. 10–11.

[55] Edward Wakefield to Francis Place, 7 December, 1813, British Library, Add. Mss, 35, 152, f. 14.

[56] Thomas Bernard, *Pleasure and Pain, 1780–1818*, ed. J. Bernard Baker, London, 1930, p. 55.

[57] From 'The County', serialized in *Cornhill Magazine*, xii, n.s., 1889, p. 100. Begging letter writers must have found *The Charitable Ten Thousand*, London, 1896, a most helpful guide.

[58] Mrs Maddocks, *The Female Missionary Advocate*, London, 1827, p. 11.

More went further and called them characteristic British women, modest and compassionate. They were also the ones who did much of the humdrum work of the charitable societies, as we shall see. If for many women philanthropy was little more than a celebration of property, for these others it was a heartfelt duty, a duty expressed in:

> Those thousand decencies which daily flow,
> From all her words and actions.[59]

The precise number of single women who subscribed to the various charities is difficult to determine and easy to understate, because of the custom common in the nineteenth century of using Mrs in later life. There is, however, ample evidence that a vast number of unmarried women contributed to the charitable societies and that some of them saw their philanthropic donations and activities as a means of offsetting the prejudice against their marital status.[60] As it has been noted from the aforementioned study of wills, there is reason to believe that single women were so forthcoming because they often had fewer family obligations than married women. There is also reason to suggest that single women often saw the objects of their philanthropy in terms of their family. Without 'natural duties' to engage them, they were advised to take up 'positive duties', among which those of a charitable nature took precedence.[61]

It is usually difficult to determine the social class of the women from the limited information included in the subscription lists, but the nature and the location of a charity often give an indication. It is likely that the majority of female subscribers were middle class, using the term in its widest sense to include all those who came between manual workers earning wages on the one hand and the landed gentry on the other.[62] Titled ladies contributed in considerable

[59] The lines are Milton's, quoted in Hannah More, *Strictures on the Modern System of Female Education*, 2 vols., London, 1799, i, p. 7.

[60] See, for example, Catharine Cappe, *Observations on Charity Schools, Female Friendly Societies, and other Subjects connected with the Views of the Ladies' Committee*, York, 1805, p. x.

[61] Miss E. J. Whately, *Cousin Mabel's Experiences: Sketches of Religious Life in England*, London [1870], p. 105.

[62] See G. Kitson Clark, *The Making of Victorian England*, London, 1962, p. 5.

numbers and were especially prominent, in both men's and women's societies, as patronesses. From the evidence of the subscription lists there is little reason to suggest that untitled women were increasing their influence at the expense of the titled to any great extent, though as the century progressed and the number of contributors vastly increased the untitled must have taken over a larger share of the contributions. The number of titled women varied from society to society of course, depending on their affiliations. In the Religious Tract Society, Nonconformist in origin, very few titled female subscribers appeared. In the Royal Jennerian Society, on the other hand, as many as two-thirds of the subscribers were titled in some years. In many other institutions, especially those with royal patronage, roughly 5 to 15 per cent of the women contributing were titled.

In any study of organized, public charity the contribution of working-class women is likely to be underplayed, for so much of the philanthropy of the poor to the poor was informal. Indeed, such a study will also understate the extent of even middle-class charity, since such a great deal of the time and energy expended on good works by better-off women was accomplished informally within the family or around the neighbourhood. Propping up family institutions and the local community were intimately associated with a woman's role, whatever her class. The philanthropy of working-class women was typically casual: dropping in on friends in distress, providing Sunday dinners for deprived children, giving free lodging or a reduction in the rent to hard-up neighbours or tenants, finding employment for friends and relatives, taking in the washing or cooking for families under the strain of illness, or simply dropping a coin in a hat passed around in a pub to support someone who might have lost a purse or been transported to Australia. Such activities were part of the day-to-day, unadministered lives of the poor; for the historian they rarely leave a trace behind.[63]

[63] The following works suggest the variety of informal working-class charity: Conybeare, *Charity of Poor to Poor; Memoir of Kitty Wilkinson of Liverpool, 1786–1860*, ed. Herbert R. Rathbone, Liverpool, 1927; J. Deane Hilton, *Marie Hilton, Her Life and Work 1821–1896*, London, 1897; C. J. Montague, *Sixty*

Donations by working-class women to institutional charity were not uncommon, however, and they can be detected in the subscription lists of various charities under such headings as 'a poor woman's mite', 'a servant girl', or 'a labouring woman'.[64] Some of them subscribed through female trade unions to such projects as the setting-up of infant schools.[65] Servants were known to found their own charities, to aid fellow servants in distress or to promote religious causes.[66] As we shall see, there were working-class Bible societies, missionary associations, and visiting charities in which labouring women were important contributors. In short, few forms of charitable enterprise went without their support. And while it is impossible to put a percentage on their numbers, their financial contribution did not go unnoticed or unappreciated by their social superiors. 'Poor contributions', announced *The Christian Mother's Magazine*, 'whether we consider the proportion which they bear to the whole wealth of the givers, or their aggregate amount are, in effect, beyond all comparison the most important'.[67] *The Evangelical Magazine*, with reference to one form of philanthropic activity, agreed: 'The pecuniary offerings of the pious poor, both with respect to their aggregate amount and the sacrifices they involve, ought to be regarded as the most precious portion of the funds raised for the spread of the gospel among the heathen.'[68]

It is open to question whether the poor wished to make such sacrifices. If one were to accept the official pronouncements

Years in Waifdom, London, 1904, pp. 125–6; Mary Pryor Hack, *Self-Surrender*, London, 1882, pp. 215–40; Robert Gregory, *The Difficulties and the Organization of a Poor Metropolitan Parish*, London, 1866; *Chambers's Miscellany of Useful and Entertaining Tracts*, iii, 'Annals of the Poor: Instances of Female Industry and Intrepidity', Edinburgh, 1845; Mrs Elizabeth Garnett, *Our Navvies: A Dozen Years Ago and To-day*, London, 1885; The Revd. E. A. Hoare, *Notable Workers in Humble Life*, London, 1887. See also Harrison, 'Philanthropy and the Victorians', pp. 368–9.

[64] I underestimated the importance of working-class contributions in 'Women in English Philanthropy, 1790–1830', *IRSH.*

[65] *Pioneer*, No. 34, April 26, 1834.

[66] See, for example the Servants' Institution in appendix II, and Dudley, *An Analysis of the system of the Bible Society*, p. 355.

[67] *The Christian Mother's Magazine*, ii, October, 1845, p. 640.

[68] *The Evangelical Magazine*, xxix, n.s., April 1851, p. 226.

of the societies there would be little doubt that the labouring classes contributed without hesitation and without harm to their standard of living. But critics sometimes challenged such assertions with the argument that indiscriminate canvassing compromised and impoverished the working classes. Servants, for example, were not completely free to turn a child or lady collector from the door. Could they hold back their pennies without alarming their employers? One gentleman, disturbed by the aggressive collectors of the British and Foreign Bible Society, contended that the poor, 'in many instances', gave because of the 'solicitations' of their employers and not because of compassion for the objects of charity.[69] The penny a week could be a 'cruel and unnecessary tax upon the hard-earnings of the lower classes' wrote another critic of the Bible Society's visiting system.[70] Everything was not quite right, or agents of the Society would not have contacted pawnbrokers to discover whether or not their subscribers had hocked their bibles and tracts.[71] There is, however, ample evidence that many poor people worked on behalf of charitable institutions and that many others saw the agents of charitable societies in more flattering terms than excise officers. Most of the working women who contributed to organized philanthropy probably did so willingly; such acts sprang naturally from their traditions and aspirations.

The more closely we look at charitable fund-raising, whether organized or informal, the more we detect women at work behind the scenes. In managing committees, sub-committees, auxiliaries, or as casual helpers, they were full of ingenious schemes. They passed the plate around at mothers' meetings, tea meetings, prayer meetings, working parties, entertainments, and exhibitions of needlework. Mrs Gladstone, to the dismay of some of her house guests,

[69] Bodleian Library, William Conybeare to Mrs Hodge, n.d., Montagu Mss, d. 12, II, f. 131. See also Martin, 'The Pan-Evangelical Impulse in Britain 1795–1830', p. 224.

[70] *Bible Associations Exposed. Being a Review of the Fourth Annual Report, of the Committee of the Henley Bible Society*, p. 15.

[71] Lloyd, *Strictures on a recent publication, entitled 'The Church her own Enemy'*, p. 128.

was known to pass it around with the breakfast dishes.[72] They contributed proceeds from pamphlets, poems, hymns, and novels; they set out collection boxes and passed around collection cards; some, like Edith Cavell in her youth, painted birthday and Christmas cards for charity.[73] Others peddled needlework bags and work baskets from door to door; a few of them even took out hawkers' licences to enable them to sell the produce of their benevolent industry.[74] Some set up sales-rooms and shops to unload their merchandise.[75] And in emergencies it was not unknown for them to sell their own jewellery and china as well. The Children's Home in London managed by Charlotte Sharmon was saved more than once by the timely sale of such items.[76]

Such activities helped to fill the coffers of the charitable societies, and they suggest that women had a considerable flair for organization. In the rapidly expanding female associations and independent women's societies, in schools, churches, and homes around the country the message had spread: influence flowed from organization. However well intentioned a woman might be, she was powerless without knowledge and connection. Thus more and more women turned their attention to accounting and secretarial skills and joined together in the common cause. Eventually they would be able to take the appropriate vocational courses in schools and evening classes, but most of what they picked up

[72] Lord Stanley, present at one of these breakfasts, remarked: 'a silly habit, which brings her into ridicule. I gave my share.' See *Disraeli, Derby and the Conservative Party: Journals and Memoirs of Edward Henry, Lord Stanley, 1849–69*, ed. John Vincent, Hassocks, 1978, pp. 219–20.

[73] A. E. Clark-Kennedy, *Edith Cavell*, London, 1965, p. 20. The auxiliary subscription lists in charities like the Church Missionary Society and the Society for the Relief of Distressed Widows provide examples of various female fund-raising methods. The Place Collection in the British Library contains an invitation ticket to an exhibition of needlework in Leicester Square in 1812, held for the British and Foreign School Society, no. 60, 1811–21. Charlotte Elliott, *The Invalid's Hymn-Book*, Dublin, 1834 and Kenney, *Charity: a poem*, are examples of books written in aid of charity.

[74] G. G. Findley and W. W. Holdsworth, *The History of the Wesleyan Methodist Missionary Society*, 5 vols., London, 1922, iv, p. 18.

[75] *Second Annual Report of the Association for the Sale of Work by Ladies of Limited Means*, [London, 1859], p. 5.

[76] Marguerite Williams, *Charlotte Sharmon: The Romance of a Great Faith*, London [1931], p. 103.

came from trial and error or from experienced women eager to pass on their knowledge. So jealous were they of their own need that in their charities they commonly refused to hire male officials, who it was sometimes said were too experienced to be trusted with the funds in any case.[77] There was more than a touch of adventure and mutual admiration in their entry into the world of organized philanthropy, and a feeling that here was a great tradition to which they might contribute if only they could master its requirements.

As the nineteenth-century philanthropic establishment discovered, often to its surprise but not always to its delight, women were mastering the required skills. They became an increasingly powerful force in the charitable world not simply because they were paying for a larger and larger share of the bill; their organizational talent and ingenious money-making also played a crucial part. More will be said later about some of their schemes for the relief of the poor and the rescue of prostitutes. The following chapters will concentrate on two of their more successful fund-raising activities, the bazaar and the recruitment of children to the cause of Christian missions. While they do not pretend to exhaust the subject of women as fund-raisers (much might be said about the charity ball and the charity shop for example), they should help us better to appreciate how women came to influence particular institutions and wider philanthropic aims.

[77] Jessie Boucherett, *Hints on Self-Help; a book for young women*, London, 1863, p. 104.

II

Bazaars

Ladies and Gentlemen, I have the honour to announce a sale of many interesting, beautiful, rare, quaint, comical and necessary articles. Here you will find objects of taste, such as Babies' Shoes, Children's Petticoats, and Shetland Wool Cravats; objects of general usefulness, such as Tea-cosies, Bangles, Brahmin Beads, and Madras Baskets; and objects of imperious necessity, such as Pen-wipers, Indian Figures carefully repaired with glue, and Sealed Envelopes, containing a surprise. And all this is not to be sold by your common Shopkeepers, intent on small and legitimate profits, but by Ladies and Gentlemen, who would as soon think of picking your pocket of a cotton handkerchief, as of selling a single one of these many interesting, beautiful, rare, quaint, comical and necessary articles at less than twice its market value.[1]

Spoken amidst trumpet flourishes by Robert Louis Stevenson's 'allegorical tout', these words introduce an institution familiar to all of us and one full of interest to the social historian—the charity bazaar. Tea-cosies, bangles, Brahmin beads, and Madras baskets may seem only quaint and comical, but they and similar trifles filled countless stalls in innumerable bazaars and raised tens of millions of pounds in nineteenth-century England for causes of every conceivable description. Men and women of all social classes found bazaars, fancy fairs, fancy sales, or ladies' sales as they were variously called, a most popular and fashionable way of making money for the charity of their choice.[2] Many philanthropic societies depended on them for annual funds. Clergymen of all persuasions, not without a touch of compromise, looked to them as a last resort to build a church or to enlarge a school or drawing-room. Managers of hospitals, dispensaries, orphanages, and asylums, coaxed by a ladies'

[1] [Robert Louis Stevenson], *The Charity Bazaar: An Allegorical Dialogue*, place of publication not given [1868], p. 1.

[2] Jumble sales and rummage sales begin to appear late in the nineteenth century; they tended to be inferior bazaars, dealing in miscellaneous, often second-hand, goods.

committee, welcomed them as a means of offsetting debts or building a new wing. Annuity companies exploited them to pension off people in the professions.[3] Virtually every other English institution, from mechanics' institutes and working men's clubs to philosophical societies and cricket clubs found that bazaars had their uses.[4] And occasionally they were even used to aid a private individual in reduced circumstances.[5] They were a marvellous invention, part amusement, part commerce, which, in Stevenson's words, made 'the exercise of charity entertaining in itself'.[6]

The Oxford English Dictionary defines a bazaar as an oriental market-place, 'consisting of ranges of shops or stalls, where all kinds of merchandise are offered for sale'. Hakluyt mentioned a 'Bazarro for marchants' in his *Voyages*; and from the seventeenth century English authors used the word in its various spellings to describe an eastern public market. One early notice of an English commercial bazaar is in Macaulay's *History of England.* Writing of life in Tunbridge Wells in 1685, he remarks: 'Milliners, toymen, and jewellers came down from London and opened a bazaar under the trees'.[7] He goes on to tell us that in 1685 a subscription was raised in Tunbridge Wells to build a church. Here we have the two traditions, the commercial and the charitable, side by side. But they do not appear to have merged until the 1790s and, more significantly, the early nineteenth century.[8]

The commercial bazaar became quite popular in the 1820s,

[3] *The Royal Lady's Magazine, and Archives of the Court of St. James's*, ii, July 1831, p. 6; *The Times*, May 3, 1850.

[4] A Christian Poet [John Holland], *The Bazaar; or Money and the Church* Sheffield [1830?], p. 34; *Leeds Mercury*, April 15, 1875; The *Newcastle Daily Journal*, March 16, 1850; March 19, 1875; The *Hampstead Record*, May 18, 1895.

[5] The *Cambridge Independent Press*, May 18, 1850.

[6] [Stevenson], *The Charity Bazaar*, p. 2.

[7] *The Works of Lord Macaulay*, 12 vols., London, 1903, i, p. 363.

[8] Something resembling a charity bazaar is mentioned in 'Frogmore Fête', a poem by Peter Pindar. He refers to some booths, filled with trinkets bought in Windsor shops, set up for charity during a festival at Frogmore [1794]. He does not use the term bazaar or fancy sale, and it appears that the booths were not central to the purpose of the fête. See *The Works of Peter Pindar*, 4 vols., London, 1794–6, iv, p. 59.

an urban variant of the English market-place. The Soho Bazaar, established in Soho Square in 1815 for the sale of fancy goods, was the first permanent London bazaar and consisted of several rooms containing two hundred stalls, fitted out with red cloth and mahogany.[9] Its novelty soon attracted large numbers of shoppers and encouraged the creation of competitors. By 1830 Londoners could choose from several, which catered to the gay and rich and sold a miscellany of 'useful and fancy articles' from pen-wipers to horses.[10] Unlike the oriental bazaar, which often took up several streets or a quarter of the city, the English counterpart was normally confined to a single building,[11] in which stalls were let to people, usually women, who plied their wares in competition with local shopkeepers and, by 1830, with the persuasive females behind the stalls of charity bazaars.

The charity bazaar came into prominence at roughly the same time as its commercial rival. In 1804 the committee of the Ladies' Society for the Education and Employment of the Female Poor considered how far 'by example or by influence' they could promote 'the use and sale of those articles, which may be manufactured by cottagers' wives and daughters at home'.[12] Unfortunately, the reports of this Society, patronized by Queen Charlotte and dominated by a host of leading evangelical ladies, do not tell us whether the idea was acted upon. But another female charity with a similar social make-up took up a variation of the idea in 1813. In that year the Ladies' Royal Benevolent Society for Visiting, Relieving, and Investigating the Condition of the Poor held a 'sale of fancy work' in aid of the charity that raised £38 in London. The sale became an annual affair, which in 1817 made over

[9] Samuel Leigh, *New Picture of London*, London, 1830, p. 213. See also Anon., *A Visit to the Bazaar*, London, 1818.

[10] *The Times*, May 30, 1829; Jan. 19, Feb. 8, June 16, Aug. 5, 1830. Commercial bazaars were popular outside London as well; see, for example, the *Brighton Herald*, March 25, 1825; *Bristol Gazette*, March 27, 1828.

[11] *Court Magazine and Belle Assemblée*, vii, Nov. 1835, pp. 178–9.

[12] *Extract from an Account of the Ladies' Society for the Education and Employment of the Female Poor*, London, 1804, appendix, pp. 16–17.

£100.[13] It continued into the 1820s, but receipts are unavailable after 1817.[14]

Such small sums imply that the 'fancy' sales of the Ladies' Royal Benevolent Society were among the earliest in England. This impression is supported by a study of London and provincial newspapers, which did not report such events until the 1820s, except for an occasional advertisement or an announcement in the society page. A survey of the ladies' magazines, the religious literature, novels, memoirs, and the reports of charitable societies reinforces the view that charity bazaars did not become widespread in England until the 1820s. During these years the number of advertisements for them increased in London and cities like Brighton, Bristol, and Leeds,[15] and short articles on them began to appear. *The Times* honoured a sale of fancywork in May 1826, patronized by the Lady Mayoress, which raised over £600 to aid distressed Spitalfields weavers, and the reporter remarked that the 'ladies acquitted themselves in their *new* characters of shopkeepers most admirably'.[16]

Six hundred pounds was a considerable sum and it reflected the growth of the charity bazaar in the 1820s. But it was a figure soon dwarfed by the receipts coming in from fancy sales in London and other parts of the country. A four-day fancy fair held at The Old Ship, Brighton, in December 1827 raised £1,315 for the Sussex County Hospital.[17] A 'Ladies' Bazaar' for Spanish refugees, under the patronage of the Duke of Wellington, produced £2,000 in May 1829 at the Hanover Square Rooms in London.[18] In the same rooms in June 1831, a sale of ladies' goods, attended by the Queen, made £3,134 in its first two days for the Irish

[13] *The Ladies' Royal Benevolent Society (Late Dollar) for Visiting, Relieving, and Investigating the Condition of the Poor at their own Habitations*, London, 1818, pp. 26-30.

[14] *The Times*, April 8, 1823, April 5, 1824.

[15] Ibid., May 11, May 15, May 30, 1826; *Brighton Herald*, March 25, Aug. 2, Aug. 12, 1826; *Bristol Gazette*, March 27, Dec. 25, 1828; *Bristol Mercury*, Oct. 27, 1829; *Leeds Mercury*, Aug. 22, 1829.

[16] *The Times*, May 30, 1826 (my italics).

[17] J. G. Bishop, *'A Peep into the Past': Brighton in the Olden Time*, Brighton, 1880; p. 177, note.

[18] *The Times*, June 1, 1829.

Distress Committee.[19] The British Orphan Asylum at Kingsland received £1,460 in October 1832 from the proceeds of a bazaar in London's Egyptian Hall.[20] And a four day 'Grand Fancy Fair and Bazaar' at the Hanover Square Rooms in June 1833, which included a stall presided over by the Queen and a variety of amusements, raised £5,106 in aid of the Society of Friends of Foreigners in Distress.[21] The Queen's blessing was the final accolade; by the early 1830s the fancy fair's position among the fashionable was secure. Around the country more humble women followed their lead. Within two decades the bazaar had taken on the forms which it was to retain, with only minor modifications, for the rest of the century. Its marriage to the fair with its wealth of entertainments, which was consummated by 1830, was a long and fruitful one.

From a study of newspaper advertisements, it appears that the number of bazaars kept pace with population growth, and perhaps more to the point, with the growth of charitable institutions.[22] *The Times*, the best guide to the most fashionable fancy fairs in London, advertised six bazaars in 1830, fourteen in 1850, twenty-one in 1875, and sixteen in 1895.[23] But as a study of local newspapers in London illustrates, these bazaars were only the tip of the iceberg. *The East London News*, which served a far from wealthy area, advertised six fancy sales not in *The Times* in 1875.[24] And *The Hampstead Record* reported or advertised seventeen not mentioned in *The Times* in 1895.[25] The drop in the number of bazaars

[19] Ibid., June 20, 1831.

[20] *The Ladies' Penny Gazette*, Nov. 3, 1832, p. 9.

[21] *An Account of the Society of Friends of Foreigners in Distress . . . for the Year 1866*, London, 1866, p. 50; *The Times*, June 20, 1833.

[22] Any analysis based on numbers of newspaper advertisements is hazardous, for newspapers often changed advertising policy or rates; and many of them in the nineteenth century changed from weeklies to dailies.

[23] *The Times*, 1830: March 29, April 27, May 14, 20, June 12, July 17. 1850: Jan. 17, April 8, 17, May 10, 13, June 1, 21, July 1, 6, Aug. 13, 21, 26, Nov. 19, Dec. 24. 1875: Feb. 12, April 17, 26, May 4, 11, 18, 26, June 1, 2, 10, 18, 19, 21, 22, 25, July 2, 31, Nov. 8, 13, Dec. 13. 1895: Jan. 19, March 22, 25, 28, April 23, May 7, 8, 13, 25, 28, June 8, 10, July 2, Oct. 5. *The Times* also reported a vast number of provincial bazaars. In January 1900, for example, it noted thirteen in its subscription lists, most of them in aid of the Transvaal War Fund.

[24] The *East London News*, 1875: Jan. 1, May 21, Sept. 17, Dec. 10, 24, 31.

[25] The *Hampstead Record*, 1895: March 23, May 4, 11, 18, June 1, 15, July 13, Nov. 16, 30, Dec. 7, 14, 21.

announced in *The Times* in 1895 can probably be explained by the emergence of local newspapers which took over some of the advertising business. During most of the century there must have been well over a hundred charity bazaars a year advertised in Greater London.

What about the provinces? Looking through selected years of a few provincial newspapers, representing various parts of the country and widely differing social groups, it is clear that a phenomenal number of fancy sales took place outside London. In 1830, *The Brighton Herald*, which had a largely well-to-do readership, advertised only one less bazaar than *The Times*.[26] The *Leeds Mercury* referred to only one fancy fair in 1829, but in 1875 it advertised sixty-three.[27] And in *The Newcastle Daily Journal* the number of bazaars announced rose to twenty in 1875.[28] If the London newspapers advertised over one hundred charity bazaars each year during most of the century, the provincial press probably advertised over one thousand annually.

What were these bazaars in aid of? Of the fifty-seven advertised in *The Times* in the years 1830, 1850, 1875, and 1895, twelve were for the benefit of hospitals and other medical charities, eleven for schools, seven for missions, foreign and domestic, and seven for church-building funds. The remaining twenty aided a variety of causes, from the Peace Society to the Soldier's Home in Aldershot. These bazaars advertised in *The Times* were not altogether representative. Of the six fancy sales mentioned in *The East London News* in 1875, four were for children's causes. And of the seventeen reported in *The Hampstead Record* in 1895, eleven were connected with churches, six of these for building funds. Church and chapel dominated in the provincial advertisements as well. Over half of the bazaars advertised in the *Leeds Mercury* in 1875 dealt with the causes of West Riding

[26] *Brighton Herald*, 1830: Jan. 2, April 10, Sept. 11, Oct. 16, Dec. 18.

[27] *Leeds Mercury*, 1829: Aug. 22, 1875: Jan. 2, 9, 16, Feb. 4, 27, March 13, 20, 24, 25, April 10, 15, 20, 22, 24, 26, May 1, 10, 26, June 12, 19, July 2, 3, 6, 10, 13, 17, Aug. 7, 21, Sept. 4, 9, 11, 13, 30, Oct. 2, 5, 6, 9, 11, 16, 19, 23, Nov. 13, 27, Dec. 6, 11, 18, 23, 24, 28.

[28] The *Newcastle Daily Journal*, 1875: Jan. 2, Feb. 4, 20, March 19, April 6, 19, May 11, 12, 17, 21, June 11, 25, July 19, Aug. 2, 11, Nov. 23, Dec. 16, 18, 23.

Congregational, Baptist, and Primitive and New Connection Methodist chapels. And a high proportion of those in *The Newcastle Daily Journal* in 1875 related to churches, chapels, and missions. We must be cautious in making too much of these figures, however, for they do not take into account the vast number of bazaars which never appeared in the newspaper press at all.

It is clear from the reports of charitable societies that thousands of small sales did not reach the newspapers, but were announced simply by word of mouth, or by a notice on a church or chapel door, or at best in the columns of a parish magazine. This was especially likely in the countryside, where bazaars tended to be smaller, where access to newspapers was relatively limited, and where news of local events could be expected to spread quickly anyway. Most of the twenty-seven ladies' sales listed in the report of the Church Pastoral-Aid Society in 1850, for example, made less than £2, barely enough to put a notice in the newspapers.[29] And at the end of the century, the Church Missionary Society, one of scores of London-based charities that received annual funds from bazaars, raised £28,817 from 1,083 fancy sales in one year.[30] Most of these sales made such small sums that the cost of advertising in the newspapers would have been prohibitive.

Without a knowledge of the number of bazaars which went unadvertised in the newspapers and therefore not knowing what percentage of bazaars were advertised, it is impossible to give anything but a rough estimate of the amount of money that changed hands across the charity stalls in nineteeth-century England. To complicate matters further, receipts are often unavailable for the bazaars that were advertised; this information can only be gleaned from newspaper accounts after the event; the reports of charitable societies, when such reports exist, do not distinguish the advertised

[29] *Church Pastoral-Aid Society, Report of the Committee, read at the Fifteenth Annual Meeting*, London, 1850, pp. 46–116.

[30] *Proceedings of the Church Missionary Society for Africa and the East. One-Hundred-and First Year, 1899–1900*, London, 1900, p. 35. The National Society for the Prevention of Cruelty to Children raised £24,238 for its reserve fund from bazaars in 1895–6. *The Power for the Children. Being the Report for 1895–96*. London, 1896, p. 34.

bazaars from the unadvertised. The sums already mentioned, however, indicate that bazaars contributed a vast amount to nineteenth-century charity. Based on the available receipts, those bazaars advertised in *The Times* appear rarely to have made less than £1,000. They often made much more. In 1845 the Anti-Corn Law League raised £25,000 in seventeen days; this was probably the most profitable and long-lived fancy fair of the century.[31] The typical bazaar advertised in *The Times* must have made something between £1,000 and £25,000, very probably a sum closer to the first figure. Whatever the figure, it would have to be multiplied by about 1,000, for that was approximately the total number of bazaars advertised in *The Times* between 1830 and 1900 (based on the four years sampled). To this figure we would have to add the funds coming in from bazaars held before 1830 and from the hundred or so smaller sales held each year in London, advertised in the local press but not in *The Times*. The accounts reported in *The Hampstead Record* in 1895 suggest that most of them produced between £100 and £1,000. In Greater London alone then, nineteenth-century charity bazaars must have made several million pounds.

The sums raised by provincial bazaars tended to be lower than those raised by the more fashionable London events advertised in *The Times*. Receipts could be minute, as the sales listed in the reports of the Church Pastoral-Aid Society illustrate. But others were on a grand scale, like the Manchester fancy fair held at the Theatre Royal in 1842, which raised just under £10,000 for the Anti-Corn Law League.[32] Usually those provincial bazaars advertised in the press made between £100 and £10,000, with the average closer to the lower figure. Given that there were probably well over one thousand fancy sales advertised each year in the provinces during most of the century, it is reasonable to estimate that they added many millions to the millions coming in from bazaars advertised in the London newspapers.[33] How much

[31] Norman McCord, *The Anti-Corn Law League, 1838–1846*, London, 1958, p. 161.

[32] Archibald Prentice, *History of the Anti-Corn-Law League*, 2 vols., London, 1853, i, pp. 300–1.

[33] The population in the provinces outnumbered that of Greater London by over six to one until the 1860s. See Best, *Mid-Victorian Britain, 1851–1875*, p. 7.

more those not advertised in the newspapers added is a matter for conjecture. But the grand total for all nineteenth-century charity bazaars must have been in the tens of millions of pounds.

When and where was this money made? The great bazaars advertised in *The Times* were usually held in the spring and early summer when the London season combined with the weather and the annual meetings of philanthropic societies to make a fancy fair irresistible. In the autumn and winter months the rage for bazaars dropped off in London, though distress calls from the manufacturing districts or from abroad could occasionally reverse this trend. And December was a special season for bazaars, especially among groups which advertised in local London newspapers. The provincial sales were less seasonal, but a preference for the spring and the week before and after Christmas can be detected. Most bazaars, whether in London or the depths of the countryside, took place during the week and lasted from one to three days, though they were sometimes carried over for an extra day or two. Bargain hunters wisely waited until the final day, when prices often dropped. If any items remained unsold, the lady organizers could unload them in another bazaar under the more humble title 'a sale of work'.

The most popular sites for the great London fancy fairs, at least in the early years, were the concert halls, like the Hanover Square Rooms and Willis's Rooms, or one of the banqueting rooms of that hotbed of radicalism in the early nineteenth century, the Crown and Anchor Tavern.[34] As the demands on the great halls increased, the hotels and the grounds of Regent's Park became popular, as did the Albert Hall in the 1870s.[35] Space and expense were the two most important considerations, and a hall with an agreeably charitable management often decided the issue. The smaller London bazaars were normally held in churches, schoolrooms, or the premises of the charity being aided. But Londoners might find themselves invited almost anywhere, from a philanthropic lady's garden to the Lord's Cricket

[34] *The Times*, May 14, 1830.

[35] *Vanity Fair*, xxii, July 12, 1879, p. 26; *The Indian Female Evangelist*, vii, April 1883, p. 103.

Ground.[36] The same variety existed in the provinces, where fancy sales could be found in town halls, theatres, churches, chapels, inns, mechanics' institutes, hotels, libraries, assembly rooms, working men's clubs, drawing-rooms, and even open fields.[37]

Not content with charity that ends at home, English women exported the fancy fair around the world in the nineteenth century. Notices of charity bazaars held in India, Australia, Africa, and North America can be found in the religious periodicals, most notably in those connected with missionary and anti-slavery societies.[38] When possible, these bazaars filled their stalls with articles produced in Britain and shipped by a ladies' committee responsible for the British side of the administration. Among the items in greatest demand were engravings, autographs, shawls, and the latest in lace, embroidery, and worsted work. The great anti-slavery bazaars in Boston, Rochester, and Philadelphia in the mid-nineteenth century often displayed the latest London fashions.[39] Sometimes such overseas enterprises backfired: in 1859 the Methodist mission in Sierra Leone received its fancy articles from England safely enough, and duty free, but before the bazaar could be held, the missionaries and their wives in Freetown who planned it were either dead or dying of fever.[40] Not to be put off, the ladies on the English end held the sale in the more hospitable clime of Cheltenham;[41] it is a mystery whether or not the goods sent to Africa were part of it. The problems connected with international fancy

[36] *Fifth Report of the British and Irish Ladies' Society*, London, 1828, p. 71; *The Times*, June 11, 1844.

[37] James Sherman, *The Pastor's Wife. A Memoir of Mrs. Sherman*, London, 1848, p. 40; *The Harbinger*, July 1861, p. 169; *Brighton Herald*, March 25, July 15, 1826; The *Cambridge Chronicle and University Journal*, May 18, 1850; The *Newcastle Daily Journal*, Feb. 20, May 21, 1875.

[38] See, for example, *The Anti-Slavery Reporter, The Harbinger*, and *Missionary Leaves*. See also *Hints on Ladies' Working-Parties, and Supplies for Missionary Stations*, London, 1857.

[39] *The Anti-Slavery Reporter*, i, n.s., April 1, 1846, p. 58; iii, n.s., Jan. 1, 1855, p. 99; iv, n.s., July 1, 1856, p. 168. Boston hosted the first of the American anti-slavery bazaars in 1833. It appears that charity bazaars became popular among the rich somewhat later in America than in England; Harriet Beecher Stowe remarks that the Boston bazaar was not very fashionable in its earliest years. See *The Anti-Slavery Reporter*, iii, n.s., Jan. 1, 1855, p. 99.

[40] *The Harbinger*, Feb. 1859, p. 16; Sept. 1859, pp. 209–10.

[41] Ibid., July 1861, p. 169.

sales were considerable; some British women concluded that it was easier to hold the bazaar at home and send a cheque.[42]

What accounts for the rise and phenomenal success of this unique institution? In the broadest terms the growth of the bazaar depended on the myriad charities in nineteenth-century England, from orphanages to missionary societies, which were expanding in response to the problems and opportunities associated with industrial and demographic change. Virtually every charity had an insatiable appetite for funds, which required new methods of fund-raising. The bazaar, like so many of these new methods, was pre-eminently a female affair, and was both cause and effect of the expanding influence of women in philanthropy. In short, where women played a role in a charity bazaars were likely to follow. No evidence has been found that men ever took the management and operation of a nineteenth-century bazaar into their own hands. The word bazaar was interchangeable with 'ladies' sale'. Whichever term was used, the institution was marvellously flexible, for it had few inherent religious or political implications. Its use hinged more on an institution's willingness to open its doors to women than on sectarian doctrine. Evangelical charities, for instance, which were relatively open to women's participation, were among the earliest and most enthusiastic in support of bazaars. And as with children's contributions to charity, the proceeds from bazaars were greatly stimulated by the women's auxiliary movement.

Some of the earliest fancy sales, as we have seen, were organized by female philanthropic societies, that is those (not auxiliaries) managed exclusively by women. By the end of the century there were a large number of these societies in England, many of them established during the decades when the charity bazaar began to be popular. Unfortunately, the records are very fragmentary for these charities, and we have only a glimpse of the role played in them by bazaars.

[42] *The Anti-Slavery Reporter*, vi, n.s., June 1, 1858, pp. 143–4; vii, n.s., Dec. 1, 1859, pp. 277–8.

Clearly, the Ladies' Society for the Education and Employment of the Female Poor showed an interest in something like a bazaar as early as 1804; and after 1813 the Ladies' Royal Benevolent Society depended on bazaars for roughly 10 per cent of their annual funds. In the 1820s, the evangelical British and Irish Ladies' Society also received annual funds from the sale of ladies' fancy goods.[43] It also devised an interesting variation, when it set up shops around Britain to sell the goods manufactured by the female poor;[44] this was the idea recommended by the Ladies' Society for the Education and Employment of the Female Poor. By 1826 the Society of Charitable Sisters, a Catholic institution, was appealing to the nobility and gentry to attend its annual bazaar to relieve 'poor aged widows and single women of good character'.[45] All of these female societies were well connected and drew upon the wealth and talent of ladies with a considerable knowledge of public life. It is to them, along with the countless ladies' auxiliaries, that we must look for the ideas, enthusiasm, and administrative experience which in a few years turned the charity bazaar into such a prominent institution.

The popularity of the charity bazaar must not be seen only in terms of the demands of philanthropy and the growth of women's charitable activities but also as an aspect of nineteenth-century entertainment. It was, of course, not the only recreation designed by philanthropists to lighten hearts and pockets. Charity balls, festivals, benefits, dinners, concerts, sermons, and cruises[46] were part of English life before the charity bazaar became fashionable, and they must have helped to give the bazaar a sense of tradition and respectability. The part played by ladies in organizing charity balls provided useful experience in setting up bazaars; London's Lady Mayoress, Mrs William Hale, and Lady Salisbury, to name only two from the first half of the century, actively

[43] *Fifth Report of the British and Irish Ladies' Society*, p. 71.

[44] *Third Report of the British and Irish Ladies' Society*, London, 1825, pp. v., 50–2.

[45] *The Times*, May 15, 1826; April 27, 1830.

[46] Charity cruises, or 'water parties' as they were sometimes called, travelled from London to places like Sheerness and cost the benevolent £1 or so. See *The Times*, July 13, 1830.

promoted both charity balls and bazaars.[47] But the charity ball and the other pleasures associated with philanthropy were perhaps not so admirably suited to human behaviour. Unlike the bazaar they did not encourage people to play at 'one of the dearest pleasures to the Human Heart'–shopping. They did not give the benevolent, again to use Stevenson's words, 'a direct and emphatic sense of gain'.[48] The excitement of the charity bazaar was not unlike that of its cousin the commercial bazaar, though it brought the extra satisfaction of the performance of a duty. Charles Greville captured its atmosphere in a description of the 'Grand Fancy Fair' in aid of foreigners in distress at the Hanover Square Rooms in 1833:

It was like a masquerade without masks, for everybody–men, women, and children–roved about where they would, everybody talking to everybody, and vast familiarity established between perfect strangers under the guise of barter. The Queen's stall was held by Ladies Howe and Denbigh, with her three prettiest maids of honour, Miss Bagot dressed like a soubrette and looking like an angel. They sold all sorts of trash at enormous prices.[49]

The fancy fair was an ideal afternoon entertainment for the leisured class and was seen as such by promoters, who normally advertised in the amusement sections of newspapers and magazines, and by visitors, like Greville, who could take time from his schedule to saunter through crowded galleries admiring the ladies. What amusements outside the home were available between the hours of noon and six, when a fancy fair was in full swing to the music of a military or quadrille band? In London, which offered the greatest variety, the leisured could attend a club, an afternoon concert or a horse show, Madame Tussaud's, perhaps an exhibition of needlework or water-colours. The public parks offered a diversion to anyone respectably dressed (and some who were not) and for three shillings the Colosseum in Regent's Park provided a view of the city's

[47] *The Journal of Mrs. Arbuthnot, 1820–1832*, 2 vols., London, 1950, ii, p. 26; *The Times*, May 3, 26, June 7, 1826; Oct. 25, 1832.

[48] [Stevenson], *The Charity Bazaar*, p. 2.

[49] Charles C. F. Greville, *A Journal of the Reigns of King George IV and King William IV*, 3 vols., London, 1875, ii, p. 383.

panorama.[50] Many ladies and gentlemen spent idle hours browsing through shops or one of the commercial bazaars. A charity bazaar, for an admission price of one shilling or two shillings and six pence, children half-price or free, was excellent value, especially for those in that listless state of mind who are looking for something, but nothing in particular.

The fancy fair was pleasure usefully channelled, and as far as the organizers were concerned, the greater the pleasure the greater the usefulness. They spent every effort devising new attractions and entertainments as part of the bazaars, and by the 1840s some of them resembled a carnival more than a market. It could be argued that no other entertainment, not even a carnival, offered such variety. Women showed an eclectic taste, and with at least one eye on effect they spared nothing and no one. Palmists and character readers, wild animals, poets, actors, Christy Minstrels, magicians, lady harpists and guitarists, the Queen, children in fancy dress, soldiers and sailors in formation, and just about anyone or anything else available might confront the bemused visitor to a bazaar.[51] The stallholders themselves might be found wearing the costumes of Dickens's characters and selling anything from flowers and machinery, to a portrait of Pius IX.[52] The annual fancy fair of the Dramatic College in Woking regaled the visitor with a burlesque performed by the Savage Club and invited a 'brilliant galaxy' of stars from the theatre world to act as stallholders.[53] The Royal Ear Dispensary called its fancy fair in Regent's Park a 'Fête Champêtre and Ladies' Sale' and paid two English military bands, a French band, and a troupe of Hungarian singers to perform. 'One of

[50] *The Times*, Feb. 15, 1830; G. W. Thornbury and Edward Walford, *Old and New London*, 6 vols., London [1873 etc.], v, pp. 269–71.

[51] *A Guide to Bazaars and Fancy Fairs, the Organisation and Management, with Details of the Various Devices for Extracting Money from the Visitors*, London [1882]; *The Times*, July 6, 1850; March 28, April 23, June 27, 28, 1895; *Leeds Mercury*, July 3, 1875.

[52] The *Chelsea News and General Advertiser*, July 29, 1865; *National Anti-Corn-Law League Bazaar Gazette*, number 3; *Woman's Mission*, ed. Burdett-Coutts, p. 78.

[53] *The Ladies' Cabinet of Fashion*, xix, July 1861, p. 53.

those foreign species of non-descript amusements', remarked an English critic.[54] The Licensed Victuallers, not to be outdone, put on their 'Grand Fête and Fancy Fair' at the Royal Surrey Zoological Gardens and included for an admission fee of one shilling, not only ancient and modern music, but also a menagerie and a fireworks display that concluded with a 'collossal transparency of Napoleon on horseback'.[55]

Considerable thought went into the opening ceremonies. Speeches, songs, and royalty were *de rigueur* on grand occasions, but most bazaars had distinctive little touches. Those in aid of hospitals laid on nurses in their picturesque blue uniforms to look conspicuous.[56] Those in aid of the welfare of children got the children to perform.[57] Those in aid of religious causes had to settle for a cleric, to open and close the proceedings with a prayer.[58] Prayers normally opened the bazaars organized and run by children.

Children's bazaars, as we shall see, often focused on a particular theme, most notably the plight of the heathen. The great fancy fairs held in London and the large provincial cities were also often topical, but unlike the children's bazaars they taxed the ingenuity and industry of scores of women and required the setting-up of ladies' committees to organize them. The 'Press Bazaar', which took place in the Hotel Cecil on June 28–9, 1898, to aid the London Hospital, created distinct problems. The brain-child of Mrs J. A. Spender, the Press Bazaar had to recruit the patronage of newspapers and magazines and to enlist a collection of distinguished stallholders. To reflect its theme, the ladies set up a post office to handle messages in the Hotel and brought in a printing press to publish the *Press Bazaar News* on the spot. This worthy little sheet, the highest priced daily in the world at one shilling during its brief life, ran the motto: 'Find a Duchess, flatter her, and get £500.' It also spread the rumour that Joseph Chamberlain had purchased a bust of Kruger and

[54] *The Englishwoman*, number 31, May 30, 1835, p. 483; *The Times* May 21, 1835.
[55] *The Times*, July 6, 1850.
[56] *Press Bazaar News*, June 28, 29, 1898.
[57] *Brighton Herald*, Sept. 18, 1830; *The Times*, June 27, 1895; Marie Hilton, *The Fourth Year of the Crèche*, London [1875], p. 68.
[58] *Missionary Leaves*, i, Aug. 1, 1868, p. 58.

that Sir William Harcourt was seen leaving the hall with a volume of Foxe's *Book of Martyrs.*[59] Such little touches amused the guests and opened their purses. To the satisfaction of Mrs Spender and the London Hospital, the sales added up to £10,000.[60]

The Anti-Corn Law League Bazaar in Covent Garden Theatre, which ran from May 8 to 27, 1845, posed an even greater challenge to the ladies. They shipped goods from all over England into Covent Garden—needlework, carpets, shawls, and muslin from Rochdale, Halifax, Leeds, and Bradford; lace and hosiery from Nottingham and Leicester; hardware and fancy goods from Manchester and Birmingham; machinery and tools from Sheffield: twenty-seven stalls in all, representing the talent and industry of every class in British life, from working men to fashionable ladies. Special trains were laid on to bring goods and passengers to London. A special paper, *The National Anti-Corn-Law League Bazaar Gazette*, published daily in the Theatre, informed the guests about events, sang the praises of free trade, and punned atrociously. Eight to nine thousand people had to be accommodated each day from noon to 10 p.m., fifteen hundred of them present at any one time. Like the Press Bazaar it contained a post office and sold a strange miscellany of items, a poem on free trade by Horatio Smith, a collection of autograph letters by George Washington sent from Leeds, and a shoe made without seam or stitch by an Ashton craftsman. From the 'Lost and Found Department' the message went out: 'Lost, the cause of the monopolists'.[61]

It was quite a show, part amusement, part trade fair, and part politics, a magnificent tribute to the zeal of free traders, the claims of British industry, and 'the intelligence, tact, and spirit of self-sacrifice' manifested by the ladies who suggested

[59] *Press Bazaar News*, June 28, 29, 1898.

[60] ibid.; *Daily Mail*, June 29, 30, 1898; The *Sketch*, June 29, July 6, 1898; *The Times*, June 29, 1898.

[61] *National Anti-Corn-Law League Bazaar, to be held in Covent Garden Theatre, London, May, 1845* [1844], pp. 1–4; Prentice, *History of the Anti-Corn-Law League*, ii, pp. 327–41; *The Poetical Works of Horace Smith*, 2 vols., London, 1846, i, pp. 217–19; Anon. [William Atkinson Gardner], *A Rythmical Notice of the Anti-Corn-Law League Bazaar*, [London, 1845]; *National Anti-Corn-Law League Bazaar Gazette*, numbers 1–17.

the idea and followed it through to completion.[62] In the end it cost £5,713 to put on and netted £25,000.[63] It was a preview of the Great Exhibition. It also ennobled the bazaar as an institution, announced *Douglas Jerrold's Magazine*, turned 'a commonplace thing—often an idle mart for children's trumpery' into 'a great and holy thing'.[64] High praise to lavish on a fancy fair, but when shorn of the hyperbole of a free trader on the loose, it reminds us of the bazaar's great flexibility as an agency for raising money. Without inherent political content, it could be turned to any end, a fact recognized in the twentieth century by the suffragettes.[65] And without permanent bricks and mortar it could be set up anywhere, for purposes great or small, religious or secular.

Such flexibility had its disadvantages, most notably that each fancy fair had to be set up from scratch. As the Press Bazaar and the Anti-Corn Law League Bazaars illustrate, this could be a sizeable chore, and one that fell heavily on the women who worked on and off for weeks, sometimes months, behind the scenes. In the largest bazaars ladies' committees had to be formed not only in the host city but in cities and towns everywhere. The Ladies' Committee of the Manchester Anti-Corn Law League Bazaar, which worked closely with the Anti-Corn-Law and Anti-Monopoly Associations, ultimately contained 360 members from all parts of the kingdom, each lady acting as a link with her particular community.[66] Friends and relatives had to be solicited for funds and articles, and an impressive amount of administration done to make sure that the contributions reached the Theatre Royal on time. An interesting twist was given to such activities in 1845 by the Manchester Ladies' Committee which worked on behalf of the London Anti-Corn Law

[62] Prentice, *History of the Anti-Corn-Law League*, ii, p. 336.

[63] *National Anti-Corn-Law-League Quarter of a Million Fund*, Manchester, 1846.

[64] Quoted in Prentice, *History of the Anti-Corn-Law League*, ii, p. 339.

[65] It is not very well known just how indebted the women's suffrage movement was to the bazaar, but *The Suffragette* gives us a clue. In the issue of November 8, 1912, for example, local unions reported over twenty jumble sales or bazaars.

[66] Prentice, *History of the Anti-Corn-Law League*, i, p. 298.

League Bazaar. Mrs Woolley, who was active in the Manchester bazaar three years before, suggested that Manchester should be divided into districts and that women should canvass for contributions.[67] Here was the system of district visiting applied to the purposes of a bazaar. It was an idea that perhaps could only be devised and carried out by women who had the district-visiting experience and the leisure.

But this was only the beginning. Once the date was set the hall had to be prepared, the catering looked after, tickets printed and sold, transport considered, opening celebrations planned, patrons and stallholders recruited, and advertising arranged. (Advertisements for the ambitious bazaars were often inserted in newspapers of other cities.)[68] Around the country 'Dorcas needles'[69] were put to work, pincushions stuffed, dolls dressed, and odds and ends from former bazaars pulled out of drawers to be freshened up for sale. The day before the opening the placards would be designed, pasted, and posted, tables set up, decorated, and covered with wares. 'How joyfully the young ladies assemble the day before the bazaar to do the work of decoration, and how fagged they are before evening, how weary of the sight of pink and blue glazed calico and laurel branches and paper flowers, of hammers and of nails.'[70] It could be an exhausting business, which kept women in a state of frantic expectancy, but most of them gloried in the challenge. Like Marie Hilton, who ran a bazaar every two years in aid of her Crèche on the Commercial Road in the East End of London, they welcomed the opportunity to make arrangements and meet old friends.[71] Musing on her preparations for 'bazaar week' and the prospect of 'standing' from

[67] Ibid., ii, p. 315.

[68] *Brighton Herald*, May 11, 1850; *Chelsea and Pimlico Advertiser*, July 29, 1865; The *Hammersmith Advertiser*, July 15, 1865; *Leeds Mercury*, Aug. 22, 1829.

[69] From the biblical woman Dorcas (Acts 9: 36) who made clothes for the poor. The name became a brand of needle and was also applied to meetings, which had as their purpose the making of garments for the poor. See *The Needle, A Magazine of Ornamental Work*, ii, Jan. 1 [1853], advertisement, back cover.

[70] *The Ladies' Cabinet of Fashion*, xxx, Feb. 1867, p. 94.

[71] Hilton, *Marie Hilton*, p. 291.

10 a.m. to 5 p.m., Elizabeth Gaskell exclaimed: 'But Oh what teas we must will have.'[72]

Getting patrons and 'standers', as stallholders were sometimes called, was a problem, but one worth taking trouble over. A stall presided over by the Queen gave the greatest mark of distinction and guaranteed a large turn-out of nobility and gentry, and a tidy profit. Consequently, requests for the Queen's patronage bombarded the royal household from all directions. The Leeds Mechanics' Institution, for one, tried repeatedly to get Queen Victoria to favour their annual bazaar.[73] Failing the Queen any distinguished lady, preferably with a title, ensured an increase in sales. And as some organizers discovered to their delight, today's standers could be tomorrow's benefactors.[74] If a duchess could not be enticed, a famous actress would do very nicely. Sarah Bernhardt and Ellen Terry spent their share of time behind a stall mobbed by autograph seekers.[75] Or failing an actress, the wife of an MP or an eminent cleric would do. Such luminaries were not always to be found and were not always sought. Working-class girls took their places behind the stalls of the annual bazaar of the London Girls' Club, founded by Flora Freeman in the late nineteenth century.[76] This was their affair, organized and carried out by the Girls' Club Committee; it is unlikely that a duchess would have been invited to take a stall. Other fancy sales would be a happy mixture of classes—the lady of the manor might be found selling antimacassars next to a labourer's wife selling flowers, or an artisan's wife might be in the stall next to the wife of a middle-class patron at a sale hosted by a mechanics' institute.[77] In such small ways the two nations were brought into contact.

In most charity bazaars the promoters had to settle for the

[72] *The Letters of Mrs. Gaskell*, eds. J. A. V. Chapple and Arthur Pollard, Manchester, 1966, p. 13.

[73] Public Record Office, HO 45 0S8064/1.

[74] Frances Martin, *Elizabeth Gilbert and Her Work for the Blind*, London, 1887, p. 223.

[75] *Daily Mail*, June 30, 1898; *Vanity Fair*, xxii, July 12, 1879, p. 26.

[76] Flora Lucy Freeman, *Religious and Social Work amongst Girls*, London, 1901, pp. 44, 142.

[77] A Stallholder, *A Fancy Sale*, London [1875?], p. 11.

assistance of local women, friends or relatives, perhaps members of a maternal association or a prayer group, who both stitched and sold. For more humble women like these, who were after all in the majority, the annual bazaar could easily be fitted into the domestic routine and for a few days of the year offered relief from it. It assuaged restless energy and, not without a measure of publicity, provided an innocent outlet for those pillars of female respectability, compassion and needlework. Women of modest station supervised the children's bazaars and rallied around the wives of clerics who often took matters into their own hands when a church needed repairing or a debt needed paying. In 1835 Martha Sherman, the wife of the evangelical preacher James Sherman, set her friends to sewing in Reading, pitched a tent on the Oxford Road, embellished it with flowers, prints, needlework, and prayer, and sold enough fancy goods in two days to pay off a £300 debt owed on a local chapel.[78] This, and not the splendour of Covent Garden Theatre, was what the charity bazaar meant to most women.

An atmosphere of friendly female rivalry pervaded fancy fairs. 'For conquest dressed', as a 'Younger Brother' put it,[79] stallholders battled to capture the market in their line of goods and end the day with the highest profits. And in the pursuit of a good cause, or a rich man, they did not hesitate to use feminine wiles. If they stayed behind their counters a man might have a chance, bemoaned one gentleman: 'But not content with engaging him in front, they throw out scouts; and light troops (of young ladies), in skirmishing order, are spread over the field; . . . He must buy that enormous pincushion, . . . and the baby's cap, and the box of chocolate, and put his name down in the raffle for an "old master".'[80] Nor were some standers above more dubious tricks. Running out of change was a favourite ploy, as was 'borrowing' an attractive item from a neighbouring stall and selling it at half-price.[81] One enterprising lady found the Duke of Wellington's umbrella near her stall and before he

[78] Sherman, *The Pastor's Wife*, pp. 40–1.
[79] A Younger Brother, *On Charity Bazaars*, London, 1829, p. 16.
[80] *Cornhill Magazine*, iv, 1861, p. 339.
[81] A Stallholder, *A Fancy Sale*, p. 12.

could fetch it she cried, 'who will give twenty guineas for the Duke's umbrella', and it was gone.[82] If we are to believe one stallholder turned poet, out and out dishonesty was not unknown:

> But I fairly blushed at the manner in which
> She deceived Mr. V., so good natured and rich:—
> She twisted the 5s ticket right round,
> And said the cigar case was marked at a pound![83]

Such practices would have ruined the reputation of a tradesman, but to a lady turned shopkeeper, it was all in a day's work.

Many a bachelor got more than he bargained for. The charity bazaar offered wonderful possibilities for flirtation and romance; where else could respectable 'young ladies exhibit themselves for three days, from twelve to four, for the small charge of one shilling per head'![84] In the opinion of Dickens, 'visions of admiration and matrimony' floated before the eyes of many an 'aspiring' young lady, with the result that 'some wonderfully meritorious institution, which, by the strangest accident in the world, has never been heard of before, is discovered to be in a languishing condition'.[85] *Punch* offered some 'Bazaar Rules', worked out 'on the summits of mountains and the tops of omnibuses', which included advice to young men on flirtation: 'Perhaps your wisest plan will be to go at once to Mamma, state the case fully to her, and be guided implicitly by her decision.'[86] The popular song 'I met her at a Fancy Fair'[87] was not unknown to Mamma. The romantic undertones of the charity bazaar and the emphasis given to them by writers are striking. (Did not the masculine preference for Maggie Tulliver's commodities cause a fuss among her erstwhile friends?)[88] They suggest that a great many men attended nineteenth-century bazaars and that women suffered from a lack of other outlets for their romantic inclinations. Small wonder

[82] *Punch*, i, 1841, p. 95.
[83] A Stallholder, *A Fancy Sale*, p. 13.
[84] Charles Dickens, *Sketches by 'Boz'*, 2 vols., London, 1836, i, p. 137.
[85] Ibid.
[86] *Punch*, lxviii, 1875, p. 159.
[87] [Holland], *The Bazaar*, p. 17.
[88] See the chapter 'Charity in Full Dress' from *The Mill on the Floss.*

that critics turned the fancy fair into 'vanity fair'.[89] At least one writer raised it to metaphor.[90]

Nineteenth-century society did not, on the whole, approve of women displaying themselves in public, and so it is not surprising that the fancy fair came in for its measure of ridicule and criticism. The main thrust came from the religious community. Some Christian mothers compared the vanity and frivolity of the charity bazaar with assemblies for music and refused to permit their daughters to attend.[91] Undoubtedly it was a pious mother who put up the sign 'flirting not allowed here' at one religious bazaar.[92] From as far away as Calcutta, where fancy sales were the rage in the 1850s, came a cry against this 'violation of feminine delicacy', which reduced wives and daughters to shop-girls.[93] And churchmen of all persuasions, Catholics included, could be found who warned of 'a vain-glorious inconsiderate benevolence, which is not indeed benevolence, but only a more specious form of selfishness, which is thus seeking a vent for its love of bustle and desire of influence'.[94] But while most religious writers thought a passive disposition among the finest ornaments of female character, they did not necessarily share the worry of some High Churchmen that charity bazaars undermined traditional methods of raising money; this is one reason why evangelicals found it relatively easy to come to terms with the fancy fair. According to High-Church doctrine it was the duty of clerics to encourage parishioners to bring their offerings to the altar. Instead, they were all too often endorsing a ladies' bazaar and 'the attendance of Mr. So-and So's quadrille band'.[95]

[89] See, for example, James, *Female Piety*, p. 127; *Cornhill Magazine*, iv, 1861, p. 338

[90] Anon., *The Bazaar; or Fragments of Mind*, Lancaster, 1831.

[91] *British Mothers' Magazine*. xi, Sept. 1855, p. 203.

[92] [Female Author], *A Visit to a Religious Bazaar*, London, 1857, p. 13.

[93] From *Hurkaru*, quoted in the *British Mothers' Magazine*, ix, April, 1853, p. 82.

[94] R. W. Evans, *The Bishopric of Souls*, London, 1841, p. 271. James, *Female Piety*, p. 127; Landels, *Woman's Sphere*, p. 224; Frederick William Faber, *Spiritual Conferences*, London, 1859, p. 362.

[95] J. J. Blunt, *The Duties of a Parish Priest*, London, 1856, p. 259; Owen Chadwick, *The Victorian Church*, 2 Parts, London, 1966, 1970, ii, p. 174.

Few, if any, religious writers quibbled about the use of charity bazaars in secular causes. What some of them feared, one 'Christian poet' put it directly, was that the Church was in danger of being 'womanized'.[96] Fox-hunting and dancing parsons could at least plead manliness. But the bazaaring parson, lured by a ladies' committee, descended to the 'humiliating imitation of a "barker" or "shop-walker" in a toy-dealer's saleroom'.[97] How could a cleric with a bazaar impending, 'and his own interest identified with its success or failure, say anything about the waste of time, money, or temper—the gauds and gewgaws of pride and vanity'?[98] Other critics saw such compromises turning parsons into beggars, and for what end? 'For the poor, the Church, the choir'? Too often it was to build a vicar's library or greenhouse. Was this in the spirit of him who said 'take neither scrip nor purse'?[99] Many a cleric had spasms of conscience walking through a fancy fair. And if we are to believe one rancorous critic, more than one preacher lost his eloquence, was struck 'dumb, after the fashionable experiment and pecuniary success of a Bazaar'.[100]

Nor was there a shortage of more secular criticism. As one might expect, *Punch* and *Vanity Fair* could not restrain themselves from a xenophobic outburst on occasion, and a bazaar in aid of Frenchmen in distress or a mission to the Jews came in for abuse.[101] Nor could Dickens resist the opportunity to mock the wives and daughters of tradesmen and clerks, the 'would-be aristocrats' of the middle class, who on hearing of a 'fancy fair in high life' grew 'desperately charitable' and got up some 'dingy' assembly room.[102] But perhaps the most serious charge levelled at the charity bazaar was that it came into direct competition with regular dealers in fancy goods, commercial bazaars, and not least of all with the 'necessitous females' who made useful and ornamental articles for a living.[103] This was no doubt true, but its

[96] [Holland], *The Bazaar*, p. 18. [97] Ibid., p. 19. [98] Ibid., p. 18.
[99] *The Westminster Review*, cxxxv, 1891, p. 307.
[100] [Holland], *The Bazaar*, p. 18.
[101] *Punch*, xviii, 1850, p. 168; *Vanity Fair*, xxii, July 12, 1879, p. 26.
[102] Dickens, *Sketches by 'Boz'*, pp. 136–7.
[103] *The Countess of Huntingdon's New Magazine*, Nov. 1851, p. 252; Anon., *The Fancy Fair*, London, 1833; [Holland], *The Bazaar*, p. 16.

harmful effects were to some extent offset by the sale of articles made by 'necessitous females' at fancy fairs.[104]

It could be said that bazaars not only boosted the profits of tent-sellers, quadrille bands, and publishers of needlework magazines, but also stimulated the commercial sale of fancy goods. Many women bought articles in shops to resell in charity bazaars because they did not have the time or the skill to make them at home; this was a reasonable alternative, for items bought in shops could be resold at above their market value in bazaars. Shrewd shopkeepers donated goods to fancy fairs in the hope of gaining goodwill; however much they may have grumbled in private when a ladies' sale appeared, they did not dare to object publicly.[105] Some, like John Milling of Leeds and John Mortlock & Company, Oxford Street, offered special discounts on fancy goods bought to be resold at a charity bazaar.[106] Judging from the popularity of commercial bazaars in cities like London and Brighton, the fancy sale could not have been a very damaging competitor. Both catered to the insatiable demand for fancy goods in the nineteenth century and contributed to the decorative clutter of the Victorian drawing-room.[107]

The one criticism that appears not to have been made is that of peculation. Yet the charity bazaar provided opportunities for it. Was Titus Price, the bankrupted Sunday school superintendant who pinched £50 from the Bursley Wesleyan Bazaar, simply a figment of Arnold Bennett's imagination?[108] And what was to stop the lady who 'twisted the 5s ticket right round' and took Mr V. for a pound from slipping it into her purse? Well, the nature and the organization of the fancy fair worked against such behaviour. Bazaars normally sprang from the local community, and thievery, if detected, would bring massive social disapproval. Accountants or agents of the charity took in the day's profits at the end of

[104] See, for example, *Bristol Mercury*, Jan. 12, 1850; *The Times*, April 15, 1856; *Friendly Leaves*, Nov. 1890, p. 306; A Stallholder, *A Fancy Sale*, p. 7.

[105] A Stallholder, *A Fancy Sale*, p. 10; [Holland], *The Bazaar*, p. 16.

[106] *Leeds Mercury*, Oct. 26, Nov. 19, 1875; *A Guide to Bazaars and Fancy Fairs*, advertisement opposite page 1.

[107] The fancy goods trade was big business. By 1889, imports and home production were worth about £14,000,000 a year. See *The Fancy Goods and Toy Trades Journal*, i, Feb. 2, 1891, p. 9.

[108] See *Anna of the Five Towns*, London, 1902.

each day, reducing the temptation to theft. And, of course, most of these bazaars were run by women not primarily concerned with the financial side of life; their little touches of dishonesty were for the cause. Unfortunately, at least from the historian's point of view, the charity bazaar rarely leaves any trace behind, except for a notice in the press or a mention in the reports of philanthropic societies. It was too transitory to be investigated by government authorities, such as the Charity Commissioners. And the only record of police involvement in fancy fairs comes from the stories that occur from time to time in the press, of kleptomaniacs arrested and of police being used to keep back the crowds.[109]

The crowds and the receipts told against the critics in the end. Their censure fell upon ears converted by the likes of Robert Louis Stevenson's 'allegorical tout'. Even the godly had to admit that 'large sums are frequently raised by these means',[110] so the omnivorous fancy fair swallowed up the clerics, doting mothers, and the editor of *Punch* like everybody else. What the critics failed to realize and the supporters of the bazaar only dimly perceived, was that the fancy fair was an expression of the coming of age of women in philanthropy. It suited nineteenth-century women ideally. It offered an escape from lives of refined idleness or domestic drudgery; indeed, for middle-class women it legitimized trade and manual work from which they were customarily excluded. It also provided an opportunity for public service compatible with household routine. And, not least, it was a reflection of the compassion that was thought to be at the heart of the female character. In turn, the great success of women at running bazaars gave them the practical experience and self-confidence, which, along with their other charitable activities, spurred them on to take an ever widening interest in social administration. Could women be resisted once they had such powers over the charitable purse? As one of their number, stallholder-cum-poet, reminds us:

[109] *Press Bazaar News*, June 28, 29, 1898. The standard charge per day for the hire of a policeman was 5*s*. in the 1880s. See *A Guide to Bazaars and Fancy Fairs*, pp. 2–3.

[110] *The Countess of Huntingdon's New Magazine*, Nov. 1851, p. 252.

When the ladies asserted their might
Then who was wrong and who was right?
There is not space in this history
To fathom the depths of the mystery.[111]

[111] A Stallholder, *A Fancy Sale*, p. 16.

III

Little Vessels

Where women worked children were likely to follow. Christian parents from all walks of life were anxious to give their children a religious training, and wives and mothers, whose educational role within the family was rarely challenged, took up this responsibility with enthusiasm. It was an article of faith among the Victorians, and one which was flattering to women, that the family was 'the chief educational agency' in society.[1] Christian wives and mothers, reputed for their piety, eagerly endorsed the accepted wisdom 'that religion is the indispensable foundation of family life'.[2] The beatific scene of a devoted mother reading the daily lesson or family prayers is a familiar one, and we should not forget the sense of power it must have given her. It was at a mother's knee that a child was first taught to fear the Lord and feel the burden of moral accountability. 'The very first convictions of sin in the heart', remarked one witness, 'and the desire to have it pardoned and subdued, may be dated from these seasons of maternal watchfulness and love'.[3] Such 'seasons' opened in infancy, for parents were acutely aware that many children died before their conversions could take place. Often the first lesson a mother drummed into her children was on good works, for active charity was thought to be a hopeful sign that progress towards conversion was being made.[4] Children's memoirs

[1] Walter Joseph Homan, *Children & Quakerism*, Berkeley, 1939, p. 76.

[2] Mrs [Mary] Porter, *Mary Sumner, her Life and Work*, Winchester, 1921, p. 31.

[3] T. W. Aveling, *Memorials of the Clayton Family*, London, 1867, p. 387, quoted in *Victorian Nonconformity*, eds. John Briggs and Ian Sellers, London, 1973, p. 15. See also *Extracts from the Memoir and Letters of the late Loveday Henwood*, London, 1847, p. 4; J. A. James, *The Mother's Help towards instructing her Children*, London, 1842.

[4] Christopher Anderson, *The Domestic Constitution; or, the Family Circle the source and test of national stability*, Edinburgh, 1847, p. 417; Mrs Ellis,

are full of examples of their small acts of mercy, and the concern of their mothers to point them down the straight and narrow, into what Hannah More called 'the right channel of Christian Benevolence'.[5]

As Christian women gave encouragement to the charitable energies of the young, they helped to strengthen the connection between religion and the growing recognition of the child's personality. By the early years of the nineteenth century, partly because of the decline in child mortality, children had taken a more prominent place in the English family.[6] The emergence of a literature directed specifically at children reflected this change. Much of this literature was, of course, devotional, and adults reinforced it in their homes and schools by bombarding their children with readings from Scripture. Did not the Bible, so influential in the making of the Victorian mind, give weight to the days of childhood? Josiah 'while he was yet young, he began to seek after the God of David his father'. (2 Chronicles 34: 3.) 'Suffer the little children to come unto me', said Jesus. (St. Mark 10: 14.) 'Did ye never read, out of the mouth of babes and sucklings thou hast perfected praise'? (St. Matthew 21: 16.) Mothers, ministers, and Sunday school teachers quoted and quoted again these and other biblical sayings. And as time passed they paid more and more attention to the philanthropic impulses which such an indoctrination aroused.

The range of children's charitable activity was enormous, not surprisingly perhaps when one considers the philanthropic

The Education of Character; with hints on moral training, London, 1856, pp. 93–4.

[5] See, for example, *Brief Memoirs of Remarkable Children*, 2 vols. London, 1823; *Early Religion; or, a Memoir of Sophia F. Hoare*, Birmingham [1855?]; The Revd. T. Scott, *Memoir of Mary Scott*, London [1855?]; *Her Record is on High. A Simple Memorial of M.M.T.*, London, 1855; *Hannah Orford*, London [1854]; The Revd. John Clunie, *Memoir of Miss Elizabeth Davidson*, London, 1813; The Revd. Hugh Stowell, *Narrative of the Life of Miss Sophia Leece*, Liverpool, 1820; There are several other memoirs of children cited in the bibliography. Children's fiction also commonly called on the young to be charitable; see, for example, Mrs Harriet Drummond, *Lucy Seymour; or, it is more blessed to give than to receive*, Edinburgh, 1847; Mrs Carey Brock, *Charity Helstone. A tale*, London, 1866.

[6] See Philippe Ariès, *Centuries of Childhood*, London, 1973; Lawrence Stone, 'The Massacre of the Innocents', *The New York Review of Books*, November 14, 1974, pp. 25–31.

interests of their mothers. House-to-house visiting, working parties and mothers' meetings, the League of Pity[7] and the Children's Union[8] kept many boys and girls in touch with the poor; Bands of Hope kept them in touch with the evils of drink and the cause of moral reform;[9] Bands of Mercy 'softened their manners' and taught them kindness to animals;[10] and the Children's Scripture Union reminded them of the source of their inspiration (almost all of its 700 secretaries were women in the 1880s).[11] Sunday schools, which reached a much wider section of the public than we have sometimes been led to believe, were an important agency in much of this activity, for it was often through them that charities found their way into the children's hearts and pockets.[12] The financial contributions of children can be found in virtually every type of nineteenth-century charity, from temperance to truss societies. Most charities did not wish to turn anyone away who was willing to work on behalf of national moral regeneration and consequently opened their doors to children. This was especially true of evangelical societies, for as a religion of personal experience evangelicalism put few obstacles between the young and the life of the community.

Nowhere in the charitable world did children play a more important part than in the evangelical missionary movement. The Baptist Missionary Society, the British and Foreign Bible Society, the Church Missionary Society, the London Missionary Society, and the Wesleyan Methodist

[7] For information on the League of Pity, established in 1894, see the annual reports of the National Society for the Prevention of Cruelty to Children.

[8] The Children's Union formed in 1888, in connection with the Society for Providing Homes for Waifs and Strays. See *Brothers and Sisters. A Quarterly Paper for Children.*

[9] Brian Harrison, *Drink and the Victorians*, London, 1971, pp. 192–4; See also Lilian Lewis Shiman, 'The Band of Hope Movement: Respectable Recreation for Working-Class Children', *Victorian Studies*, xvii, September, 1973, pp. 49–74.

[10] The Band of Mercy movement was founded by Mrs Smithies in 1875 and under the ladies committee of the RSPCA formed an important part of that charity's work. See the annual reports of the RSPCA.

[11] Mrs G. S. Reaney, *Our Daughters*, p. vi.

[12] Thomas Walter Laqueur, *Religion and Respectability: Sunday Schools and Working Class Culture, 1780–1850*, New Haven and London, 1976, ignores the abundant evidence for the influence of charitable societies, most notably the missionary societies, on Sunday school life.

Missionary Society were among the earliest to encourage children's participation; and they recruited, organized, and sent children into every corner of the country as collectors and tract distributors. What were the origins, nature, and effects of their work in these great charities? What did they mean to the cause and what did the cause mean to them?

In 1804 Catherine Elliott of Sheffield, aged fifteen, dropped a penny in a tin box for the purpose of giving Bibles to the poor, rallied her brother and a few friends, and became the founder of what may well have been the first children's charity in England, the Sheffield Juvenile Bible Society.[13] This little band eventually came to the attention of the British and Foreign Bible Society, which contributed to the cause by providing the children with Bibles at cost prices. About this time all of the evangelical missionary societies began to report the financial contributions of children, from both juvenile associations and Sunday schools. As early as 1808 the British and Foreign Bible Society and the Church Missionary Society received funds from Sunday school collections;[14] and in the same year the Baptists raised £30 from a juvenile society at Eagle Street, London.[15] The first regular and systematic children's association in connection with the British and Foreign Bible Society was formed in Southwark in 1812. Composed of 3,000 Sunday school scholars, it raised £2,115 in its first eight years. By 1820 children's associations from Exeter to York also sent funds to the Bible Society's headquarters.[16] The distinction of founding the first children's association in the Methodist Missionary Society goes to Mrs Anne Burton, who in 1814

[13] Thomas Bernard, 'Extract from an Account of the Juvenile Bible Society at Sheffield', *The Reports of the Society for Bettering the Condition and Increasing the Comforts of the Poor*, 6 vols., London, 1815, vi, pp. 183–9. Bernard dates the beginning of Elliott's Society in 1805. Dudley, *An Analysis of the system of the Bible Society throughout its various parts*, p. 277, quotes a letter from one of the secretaries of the Sheffield Auxiliary Bible Society, which places Elliott's foundation a year earlier. This is probably the more reliable source.

[14] *The Fourth Report of the British and Foreign Bible Society*, London, 1818, p. lxxv; Stock, *The History of the Church Missionary Society*, iii, p. 820.

[15] *Periodical Accounts relative to the Baptist Missionary Society*, London [1808], p. 499.

[16] Dudley, *An Analysis of the system of the Bible Society*, pp. 278–81.

organized the boys and girls of Ashby de la Zouch; but the first juvenile association mentioned in the annual reports was in City Road, London, two years later.[17] By this time children's associations were well established in the London Missionary Society. In the year ending April 1815, its juvenile associations and Sunday schools brought in £1,000, more than 5 per cent of the Society's total receipts for the year.[18] 'What a respectable sum is this', announced *The Evangelical Magazine*, 'redeemed in great part, from what might else have been expended in useless trifles'.[19] It was an auspicious beginning. By the end of the century children's missionary societies connected to the London-based evangelical charities numbered in the thousands, some of them with thousands of members. As we shall see, the pennies added up to millions of pounds.

What accounts for the emergence of these associations? Some of them, like Catherine Elliott's Society, appeared to be the spontaneous response of children to the plight of their benighted neighbours, but such actions were conditioned by a rigorous religious training. Mothers, as it has been suggested, were not unaware of their power to inculcate virtue and fear in their children. 'Yes, mother, that little child of yours is mainly what you choose to make it', remarked one authority.[20] By telling their sons and daughters that they were wicked, mothers expected them to be good; and goodness could be seen in their little acts of kindness to the less fortunate, not least the heathen. 'Christian matrons', proclaimed one evangelical preacher, '. . . what more laudable ambition can inspire you than a desire to be Mothers of the Missionaries, Confessors and Martyrs of Jesus? . . . Tell the missionary story to your little ones, until their young

[17] B. Baxter to Methodist Missionary Society, 8 April, 1952, Notes and Transcripts, 6, Methodist Missionary Society Archive; *The Report for the Year 1817, of the Committee for the Management of the Missions, first commenced by the Rev. John Wesley, the Rev. Dr. Coke, and Others*, London, 1817, p. 45.

[18] *The Evangelical Magazine*, xxiii, November 1815, p. 476.

[19] Ibid.

[20] [Eleanor C. Price], *Schoolboy Morality: an address to mothers*, London, 1886, p. 5; Louisa Clayton, *Loving Messages. Addresses for mothers' meetings*, London [1885]; *The British Mothers' Magazine*, i, 1845.

hearts burn, and in the spirit of those innocents who shouted Hosanna to their lowly King, they cry "Shall not we also be Missionaries of Jesus Christ." '[21]

The juvenile missionary movement was intimately tied to the growing interest of women in the cause of foreign missions. Indeed, the growth of the children's associations went hand in hand with the growth of women's auxiliaries in the missionary societies.[22] It also went hand in hand with the work of women Sunday school teachers, for children's associations were frequently attached to Sunday schools. One adult missionary explained in the *Juvenile Missionary Magazine* why he had become a collector at the age of twelve. He ascribed it to the persuasiveness of his mother, who excited his imagination with stories of female missionaries. Inspired, he pored over his maps and prayed for the conversion of the heathen; and by the time a new minister arrived in the parish and set up a children's society, he felt himself worthy to take part.[23] Such an experience was not unique. Most Christian mothers, with visions of Borioboola-Gha[24] floating before their eyes, thought the missionary association a haven from sin and an influence compatible with their teaching at home. The associations would not have survived otherwise.

Given the preparation of the ground by pious mothers, Sunday school teachers, ministers, and children themselves, it was not difficult for agents of the missionary societies to reap the rewards. In time, all of the societies had specialists in the children's field, who advised juvenile associations already established and helped to set up new ones. Joseph Blake did much of this work for the Methodist Missionary Society. A one-time Sunday school teacher and by 1840 a circuit steward, he devised a collecting and book-keeping system, adopted by the Society in 1841, which rationalized children's operations and increased their

[21] Quoted in Geoffrey Moorhouse, *The Missionaries*, London, 1973, p. 50.

[22] See, for example, Dudley, *An analysis of the system of the Bible Society*, pp. 343–513.

[23] *The Juvenile Missionary Magazine*, xviii, April, May, 1861, pp. 105–8, 128–33.

[24] The 'Telescopic Philanthropy' of Mrs Jellyby in *Bleak House*.

contributions.[25] Charles Dudley, the highly successful organizer of female associations for the British and Foreign Bible Society, was also quick to spot the potential of children's work. He believed it a 'memorable fact' that juvenile associations were the earliest auxiliaries of the Bible Society. Taking the cue given to him by children and parents, he set out to bring these associations more intimately into the structure of the Bible Society by rules and regulations recommended by committees in London.[26]

The rapid expansion in children's associations made it imperative that their operations be rationalized and co-ordinated. The parent societies drew on the regulations of early juvenile associations to make up sets of rules which they proposed for general adoption. As early as 1821 Dudley published a list of ten 'simple' and 'intelligible' rules for children's Bible associations to observe. It was a tribute to them that these rules mirrored the organization of the parent committee and no less significantly the female auxiliary committees. They included: that each member subscribe at least one penny per week; that the business of the association be conducted by committee, with an elected treasurer and secretary; that the committee meet each month on a pre-determined date, and that five members constitute a quorum; that each collector be furnished with a collecting book, for the purpose of entering names of subscribers; and that the amount received should be given to the treasurer at the monthly meeting. The whole of the contributions, after expenses, were to be turned over to the agent of the parent society.[27] The other institutions drew up similar rules and, like the Bible Society, provided the paraphernalia of organization.[28]

Anxious to increase the sums raised by children's associations and Sunday schools, the missionary societies devised methods to ensure their steady progress. To reach the widest

[25] 'Beginnings of Juvenile Missionary Associations', Notes and Transcripts, 6, Methodist Missionary Society Archive.

[26] Dudley, *An analysis of the system of the Bible Society*, pp. 276–91.

[27] Ibid., p. 283.

[28] The Baptist Missionary Society included rules recommended for Juvenile associations each year in its *Reports. The Bible Class Magazine*, i, 1848, p. 71, also reproduced 'Rules for a Juvenile Society'.

possible audience they published children's periodicals[29] and sponsored missionary addresses. In the 1880s the Church Missionary Society arranged for addresses to be delivered simultaneously in Sunday schools across the country. It also contacted board schools and the Church Sunday School Association to impress on them the urgency of the missionary cause. And then there were lantern lectures and summer gatherings of boys and girls at seaside resorts and in the grounds of the Church Missionary College.[30] To supplement his other schemes, Joseph Blake made plans for a Methodist juvenile library and monthly prayer meetings.[31] In time, all of the societies held not only prayer meetings but also great annual children's gatherings.[32]

With the years new forms and organizations breathed fresh life into the work done by children. The Baptists, for example, established the Young Men's Missionary Association in 1848, which made a systematic attempt to bring Sunday school teachers into its programme and dispatched lecturers to visit children's groups. It also proposed to set up a missionary museum, a repository of 'rejected idols' and other 'objects of curiosity from foreign parts'.[33] In 1896, because of the increasing importance of girls in the organization, its title changed to the Young People's Missionary Association.[34] In the Church Missionary Society 'junior' replaced 'juvenile' in the vocabulary late in the century, and Miss G. A. Gollock founded the Sowers Band in 1890 to extend the work of the Society among children. By the end of the century there were 540 Sowers Bands working side by side with hundreds of 'junior' associations.[35] Such innovations kept up and expanded the interest of children, who became more and more important to the missionary cause as the years passed.

[29] See pages 90–2.

[30] *The Church Missionary Intelligencer*, xv, May 1890, pp. 297–8; *Proceedings of the Church Missionary Society for Africa and the East. Ninety-Ninth Year, 1897–98*, London, 1898, p. 35.

[31] Joseph Blake, *The Day of Small Things*, Sheffield, 1868, p. 42.

[32] One of the best documented took place in 1899; see *Church Missionary Society Centenary Meeting for Boys and Girls*, London, 1899.

[33] H. L. Hemmens, *Such has been my Life*, London, 1953, p. 98; *The Juvenile Missionary Herald*, ii, n.s., June 1849, p. 92.

[34] Hemmens, *Such has been my Life*, p. 98.

[35] Stock, *The History of the Church Missionary Society*, iii, p. 664; iv, p. 522.

Before long the societies discovered that they could not do without the children's support.

It is impossible to give a precise figure for the over-all sum that children raised for missionary societies in the nineteenth century, but certainly it was a massive amount.[36] There is some reliable financial information available for children's receipts, though it must be emphasized that there is no means of determining how much of the money came from the children themselves and how much of it they collected from adults. *The Bible Class Magazine* listed the funds raised by children in several missionary societies in the years 1841–5. The Methodist Missionary Society topped the list with £16,481, followed by the London Missionary Society with £13,195.[37] Other evidence shows that the Church Missionary Society made considerable headway among the young in the 1850s and 1860s, for in 1864 children contributed nearly £6,000, most of it from associations.[38] Progress was steady in the Society, for in the financial year 1898–9, children collected over £17,000 in missionary boxes alone.[39] By 1850 the Baptist Missionary Society, which had the smallest income of the charities studied here, expected about £3,000 each year from Sunday schools and juvenile associations, which was about 15 to 20 per cent of its annual receipts.[40] Owing to the enthusiasm of organizers like J. A. Page of Yorkshire, funds from 400 children's associations came into the Bible Society's coffers by 1869. In that year 'Young Yorkshire' alone raised £1,291 for the Society.[41]

All of the charities were adept at raising money from children for specific projects. Such enterprises aroused their romantic imagination and all the publicity that could be marshalled was used to sustain it. The London Missionary

[36] Unfortunately, the receipts from juvenile associations and Sunday schools given in the annual reports and the periodical literature are difficult to deal with, a problem further complicated because their donations are often disguised under the headings of church services, bazaars, missionary boxes, working parties, etc.

[37] *The Bible Class Magazine*, i, 1848, p. 14.

[38] *A Quarterly Token for Juvenile Subscribers*, January 1865, p. 2.

[39] *Proceedings of the Church Missionary Society for Africa and the East. One-Hundred-and-First Year, 1899–1900*, London, 1900, p. 34.

[40] *The Juvenile Missionary Herald*, ii, n.s., June 1849, p. 92.

[41] *The Sixty-Sixth Report of the British and Foreign Bible Society*, London, 1870, pp. 318–19.

Society devised a very successful campaign in the 1840s to finance its ship, the John Williams, named after a missionary who had been eaten by New Hebridean Islanders. Once excited the children responded with £6,237 and the ship was built; when it need refitting they again came to the rescue.[42] During the same years the Baptist Missionary Society appealed to children on behalf of its ship, the Dove. It put Sunday school teachers and association secretaries on the alert, and even had a song, 'The Dove', composed and distributed through its *Juvenile Missionary Herald.*[43] Armed with collecting cards and missionary boxes, the little Christian soldiers invaded their districts and brought back the spoils, ensuring that the ship continued to wing its way back and forth between England and Fernando Po.[44] In later years the Baptists ran similar appeals on behalf of native teachers and their mission in Jamaica.[45]

In 1841 the Methodist Missionary Society, which was in financial difficulties, sanctioned a children's Christmas appeal;[46] it was so successful that it became an annual affair. The Society published receipts for the Sunday school contributions to this appeal, and by 1881 also listed the additional funds coming in each year from its nearly 500 juvenile associations.

Date	Christmas Offering	Juvenile Associations	Total Home Receipts	Percentage of Children's Contribution
1841	£4,721		£75,849	6%
1851	£5,159		£75,810	7%
1861	£8,133		£101,613	8%
1871	£9,143		£110,763	8%
1881	£7,617	£9,980	£117,727	15%
1891	£7,881	£11,066	£119,080	16%
1901	£8,333	£13,860	£111,579	20%[47]

[42] *The Bible Class Magazine*, i, 1848, p. 14; *The Evangelical Magazine*, xxxiv, n.s., July 1856, p. 444 and *passim.*

[43] *The Juvenile Missionary Herald*, i, April 1845, p. 84.

[44] Ibid., December 1845, p. 272 and *passim.*

[45] Ibid., i, n.s., October 1848, p. 236; iv, n.s., April, December 1851, pp. 64, 185.

[46] *The Report of the Wesleyan Methodist Missionary Society, for the Year*

[*See opposite page for n. 46 cont. and n. 47*]

By 1901, then, children contributed about 20 per cent of the total annual home receipts of the Methodist Missionary Society. This figure would be higher if we knew the amounts they gave in church services and Sunday schools throughout the year and dropped in missionary boxes not under their direct supervision. From the establishment of the first juvenile association in Ashby de la Zouch in 1814 to the end of the century, children must have raised well over a million pounds for the Methodist Missionary Society. There is little reason to believe that in the other societies the children were less generous or less well organized. Much was expected from them, or such extensive campaigns would not have been waged to extract their pennies. Their financial contribution to the nineteenth-century missionary movement was a remarkable achievement. More remarkable when one considers that, as one child said at a meeting, they did not 'desire to enter the field as the reapers of the golden harvest, but simply as gleaners, that we may gather up the fragments that remain that nothing be lost'.[48] If we add these 'fragments' to the 'golden harvest' reaped by women for the missionary societies, we have some measure of the great debt the movement owed to women and children. By the end of the century their contributions and collections were making up roughly 70 per cent of the total annual receipts of these charities.

The children in missionary work discussed in the periodical literature and depicted in engravings are usually little girls, ranging in age from about six to sixteen.[49] The subscription

ending April 1842, London, 1842, p. 5. The Methodists also ran a campaign among the children to support their missionary ship 'John Wesley' in 1846. See 'Beginnings of Juvenile Missionary Associations', Notes and Transcripts, 6, Methodist Missionary Society Archive.

[47] See the *Reports* of the Wesleyan Methodist Missionary Society for the relevant years.

[48] Blake, *The Day of Small Things*, pp. 11–12.

[49] In 1898 the Sowers Bands of the Church Missionary Society had 13,000 members, 9,800 of them girls. *Proceedings of the Church Missionary Society for Africa and the East. Ninety-Ninth Year, 1897–98*, p. 35. See the engravings in James Bolton, *The Goldon Missionary Penny, and other addresses to the young*, London, 1868, p. 25; *The Children's Missionary Meeting in Exeter Hall, on Easter Tuesday, 1842*, London, 1842, across from title-page.

lists in the annual reports also suggest that girls were more often involved than boys. This should not be very surprising, for, as we have seen, compassion and benevolence were thought to be the preserve of the female sex. Moreover, many girls could identify with the cause of missions, for their training at home often had a missionary character. Many of the associations were made up entirely of little girls, who often worked with ladies' auxiliaries to raise funds and distribute tracts. Some were boys' associations, but most were a mixture of the two sexes, with an older child, a lady, or a Sunday school teacher acting as secretary. There was support for putting the management of the associations wholly in the hands of children, and many of them were run along these lines, despite experience which showed that 'older heads' administered with greater efficiency.[50] The average age of the children was about twelve or thirteen, though societies invited them to join from 'the youngest who can work' to 'the oldest who will attend'.[51] It was not unusual for boys and girls of four or five to be active alongside young adults in their late teens.

> There is no little child too small
> To work for God;
> There is a mission for us all
> From Christ the Lord.[52]

Nor did any single class have a monopoly in this charitable field. 'The privilege of giving is open to all', went a familiar saying.[53]

Most of the missionary societies, the Baptist and Methodist perhaps more than the others, had a large enough following among the labouring classes to diminish the problems of soliciting funds in poor neighbourhoods. In such areas children from Sunday schools and missionary associations became the ideal agents for extracting the 'widow's mite' or a halfpenny from an artisan or cottager. They could

[50] *The Bible Class Magazine*, i, 1848, p. 71.

[51] *The Juvenile Missionary Magazine*, i, December 1944, p. 156.

[52] Anna Hinderer, *Seventeen Years in the Yoruba Country*, London [1877], p. 1.

[53] *The Juvenile Missionary Herald*, xvi, June 1878, p. 95.

elicit contributions when others failed, for most parents and neighbours did not wish to look ungenerous in the eyes of a child working on behalf of charity. The young collectors were not innocent of tactics. The Reverend John Burbridge of Yorkshire, who had ninety members in his juvenile society, 'almost all of the labouring class', instructed them to solicit 'not more than one halfpenny per week from any one person, but to collect one halfpenny from as many as possible'.[54] It was a sum all but the very poorest could afford. The secret of success was that everyone, young and old, rich and poor, was seen as a potential contributor. No matter how impoverished the Englishman, the heathen could be made to look more wretched.

The child collectors were never very far from advice. Sunday school teachers, association secretaries, mothers, and the missionary magazines coached, cajoled, and tried to put them into the proper spirit for their work.[55] It was, after all, not easy to get little boys and girls to knock on unknown doors. Remember the dignity of your calling they were told—'you are not a beggar'.[56] Ask cheerfully and expect generosity. Pray. Think of the guilt and misery of sin—the love of Christ. 'Be pitiful—to the heathen/Be courteous—to your subscribers.'[57] Canvass everyone. 'A penny, is a little thing/which e'en the poor man's child may fling/into the treasury of Heaven.'[58] Walk with a humble heart. Set a worthy example to your subscribers. 'Collect early, and you are sure to collect well.'[59] Above all be punctual and systematic. Such pep talks were to transform the children into 'vessels unto honour, sanctified and meet for the Master's use, and prepared unto every good work'.[60]

While system, regularity, and punctuality were guide-lines

[54] *The Sixty-Sixth Report of the British and Foreign Bible Society*, p. 318.

[55] There is an interesting description of Sunday school children being given instructions on collecting funds in Robert Tressell, *The Ragged Trousered Philanthropists*, London, 1955, pp. 182–3.

[56] *The Juvenile Missionary Herald*, xvi, June 1878, p. 95.

[57] *A Quarterly Token for Juvenile Subscribers*, October 1856, p. 8; Bolton, *The Golden Missionary Penny*, pp. 23–8.

[58] *A Quarterly Token for Juvenile Subscribers*, April 1856, no page number.

[59] *The Juvenile Missionary Herald*, i, February 1845, p. 45.

[60] 2 Tim. 2: 21, quoted in Bolton, *The Golden Missionary Penny*, p. 276.

laid down for children to follow in their missionary work, there existed considerable freedom of expression. Children soon expanded their activities and dressed up their labours with promotional gimmicks. Numerous associations adopted names like 'the rivulet', 'the twig', 'the crumb', and 'the drop'.[61] The British and Foreign Bible Society boasted of a young working-woman in Dorset, who had trained her parrot to say 'put something in the Bible-box'. Before the bird flew away seven years later, it had collected £10.[62] Children also consecrated rabbits, chickens, pigs, and other items, animal and vegetable, to the cause. They hived bees and put them into service, made useful and fancy articles out of everything from walnuts to old clothes, and decorated their own missionary boxes and collecting cards.[63] The parent societies were quick to get the message and by the 1840s standardized collecting cards and boxes and distributed them to the children's associations. The collecting card, often beautifully engraved, had space for the names of about twenty subscribers inside, and was passed around to friends and family; when not in use it could be placed conspicuously on a mantelpiece, a potential rebuke to visitors.[64] The wooden missionary boxes, attractively labelled, came in different colours and sizes, with a slot on the top and a trapdoor underneath; the societies produced them in the thousands each year at a price of about 4*d*. a piece.[65] Children, and women too, placed them in homes, servants' quarters, churches, chapels, schools, hotels, railway stations, and other public places.[66] Opening them often called for a celebration, introduced and closed by prayer.[67] There was usually cause to be festive, for considerable sums passed through these modest containers.[68]

[61] William Canton, *A History of the British and Foreign Bible Society*, 5 vols., London, 1904–10, ii, p. 148.

[62] *Gleanings for the Young*, i, n.s., January 1878, p. 8.

[63] Ibid.; *The Church Missionary Juvenile Instructor*, vi, December 1847, pp. 379–80; *A Quarterly Token for Juvenile Subscribers*, January 1866, p. 4.

[64] *The Countess of Huntingdon's New Magazine*, 1850, pp. 82–6.

[65] *Church Missionary Record*, iii, June 1832, p. 138; *The Autobiography of a Missionary Box*, London [1896]; *A Quarterly Token for Juvenile Subscribers*, April 1857, pp. 2–3.

[66] *Children's Missionary Magazine*, iii, January 1850, pp. 19–25.

[67] *A Quarterly Token for Juvenile Subscribers*, January 1858, p. 3.

[68] Ibid., March 25, 1879, p. 8; In the financial year 1898–9 the Church Missionary Society collected £43,145 from missionary boxes. *Proceedings of the*

Such activities merged the missionary movement with children's day-to-day lives, with their pastimes and recreations. So did working parties and bazaars, which whiled away hours that might otherwise have been misspent in idleness. Children's working parties resembled those run by women, and to guard against corrupting influences a mother normally superintended them. While little boys and girls busied their fingers with pinafores, plain frocks, and pen-wipers, she counted heads and recited missionary stories. Prayers and hymns broke the monotony of these gatherings, which might last all afternoon and were run on a regular basis, usually once a week in a home or schoolroom.[69] Such occasions provided clothing to be sent abroad and filled the children's bazaars with needlework. At the same time they inculcated the habits of honest industry and benevolence. By introducing the element of benevolence, distinctions between work and pleasure, between seriousness and fancy, disappeared. To make clothes for the heathen was an expression of Christian duty. Duty done selflessly, so nineteenth-century moralists argued, brought the keenest satisfaction. 'An undutiful child cannot be a happy one.'[70]

The children's bazaar, often organized in aid of missionary work, was a very popular pastime by the middle of the nineteenth century. Frequently held during the Christmas season, it added to the gaiety of the holidays and reduced the surplus of goods built up by working parties over the year. A typical children's missionary fancy sale, held some fifty miles from London in December 1882, was initiated by some little girls, who wished 'to bring happiness to the women and children of India, who are kept shut up, and know nothing of the free happy life that most English girls lead, or worse still, know nothing of the blessed Saviour'.[71] The girls counted their pennies, rallied their

Church Missionary Society for Africa and the East. One-Hundred-and-First Year, 1899–1900, p. 34.

[69] *The Church Missionary Juvenile Instructor*, ii, n.s., January 1866, pp. 7–10; *The Juvenile Missionary Magazine*, i, December 1844, pp. 155–6.

[70] John Angell James, *The Family Monitor, or a Help to Domestic Happiness*, Birmingham, 1828, p. 203.

[71] *The Indian Female Evangelist*, vii, April 1883, p. 101.

friends and elder sisters, and set to work. They decorated their schoolroom with flags and Chinese lanterns, holly and laurel, and peacocks' feathers.

The piano, covered with a large red cloth, made a capital stand for the display of drawings, terra-cotta paintings, vases and woodwork, for sale; and close beside this was a big bran pie of penny things, the charge of which was given to a little girl who was assisting the others to sell, and very happy she looked with her little bag of pennies on her arm, getting heavier and heavier every moment.[72]

The little girls dashed in and out of their stalls like rabbits in a warren and expertly coaxed eight pounds from their customers; at the end of the afternoon they retired to the drawing-room for tea and music. From London to Swan River, Australia, such scenes were repeated, and like the women they emulated, children in their 'muslin aprons and high mob caps', gave the occasions an extra twist whenever possible, like the sale of a missionary pig, or the decoration of their schoolroom with Union Jacks.[73] While selflessness rarely displayed itself on these occasions, they encouraged Christian benevolence in the young and provided a happy outlet for patriotism and needlework. They also brought in vast sums to the missionary societies.

Perhaps the most memorable childhood experiences connected with the missionary movement came during the meetings put on by the societies each year in villages and cities around the country. Children often travelled for miles to hear an address delivered by a missionary back from Africa or Asia, who could be expected to have with him the trappings of his work, maps and globes, idols used in pagan rites, and sometimes even a regenerated heathen, saved from the jaws of hell. Juvenile associations and Sunday schools relied on such meetings to fill out their membership and inspire their collectors. And the societies used them to recruit missionaries. Miss Nellie Gilbert, for example, overwhelmed by a meeting in Brighton in 1844, became a collector for the

[72] Ibid.

[73] *The Church Missionary Juvenile Instructor*, iii, n.s., March 1867, p. 47. The pig was not to be sent abroad, just the proceeds from its sale. Missionary rabbits and hens were also popular among the children. See also *Children's Fancy Work: A Guide to Amusement and Occupation for Children*, London [1882].

cause and later married a missionary to Keppel Island.[74] On such occasions, whether a missionary or an agent of one of the societies officiated, the message was clear: a missionary calculus counted pennies gained and souls lost and ended with a prayer for the speedy conversion of the world.[75]

To provide a climax to the activities of the year, all of the societies held great annual children's meetings. Among the earliest were those staged in Exeter Hall by the London Missionary Society and the Baptist Missionary Society. There, on Easter Tuesday 1842, between five and six thousand children attended the London Missionary Society's annual celebration, conducted by a distinguished array of officials and missionaries.[76] Interspersed with prayers and hymns, the missionaries, brandishing fallen idols, recounted tales of vice, atrocity, and cannibalism that must have frightened the wits out of some of the children. For effect they trotted out an Arab girl in costume, three blind Chinese girls, two of whom read from the Scriptures in brail, and the Bechuana girl Sarah Roby, a familiar face on the circuit, who had been rescued by a missionary 'from the grave where her mother and other relatives had consigned her'.[77] After these lurid scenes several speakers lamented that the Society's income was insufficient 'to meet the cries of the heathen'.[78] The usual energetic call to action followed. An awakened spirit, proclaimed one promoter, would:

> operate on millions of young and old and of every class. Your brothers and sisters will also, my dear children, be won by your earnestness, and in their earliest childhood your mothers will act like the negro woman who placed a little bit of money between the fingers of the babe she was carrying when passing the collection-plate, that it might drop it in, observing to her minister, 'We bring dem up to it.'[79]

The revival demanded self-sacrifice with mother's milk.

Like the bazaars and working parties, such occasions

[74] Jennie Chappell, *Three Brave Women*, London [1920], pp. 76–7.

[75] *The Bible Class Magazine* calculated that 36,860,000,000 heathens had died during the Christian era and that 38 still died every minute. 'Reader, what influence should these facts have upon your mind?', i, 1848, p. 15.

[76] *The Children's Missionary Meeting in Exeter Hall, on Easter Tuesday, 1842*, p. 4.

[77] *The Evangelical Magazine*, xx, n.s., May 1842, p. 247.

[78] Ibid., p. 250.

[79] Ibid.

remind us that raising money and saving heathens was only one side of the children's missionary movement. Dr John Leifchild, who chaired the Exeter Hall meeting in 1842, touched on the other in his closing address: 'When anyone attempts to benefit others, it is the benevolence of the Deity making it the source of benefit to himself.'[80] Another spokesman quoted Proverbs: 'He that watereth shall be watered also himself.'[81] From the beginning, those who promoted children's missionary work argued from a 'higher ground' and considered the movement's influence on the minds of the young.[82] 'Training' became the watchword. Christianity must be impressed on the minds of children in their earliest years; what better way to achieve this than through pious mothers, Sunday schools, and missionary associations? To many the financial consideration was secondary. 'What are all the silver and gold in the Bank of England', wrote Joseph Blake, 'compared to the intrinsic value of an immortal soul, and the means of bringing the rising generation to the knowledge of Christ and his Gospel'?[83] He might have asked whether the missionary cause itself had a future without a steady supply of devoted men and women, drawn from the children's ranks.

The high priority given to the training of children by parents and the missionary societies created an ever increasing demand for books and periodicals devoted to religious instruction. It was a demand quickly supplied by the publishers, who sought to strike a balance between what the parents would buy and what the children would read. Many of their publications, like *The Peep of Day* by F. L. Bevan, afterwards Mrs Mortimer, aimed at the infant market.[84] Others, like James Bolton's *The Missionary Stick Gatherers* and Mary Ann Barber's *Missionary Tales* were written explicitly to encourage children's missionary work.[85] Scores of

[80] Ibid., p. 251.

[81] Proverbs 11: 25, quoted in Bolton, *The Golden Missionary Penny*, p. 31.

[82] See, for example, Dudley, *An analysis of the system of the Bible Society*, p. 287; Blake, *The Day of Small Things, passim.*

[83] Blake, *The Day of Small Things*, p. 31.

[84] *The Peep of Day: or a Series of the earliest religious instruction the infant mind is capable of receiving*, London, 1870.

[85] *Missionary Stick Gatherers: an address to the members of juvenile mis-*

children's periodicals supplemented such didactic and pious tomes. All of the missionary societies printed vast numbers of children's magazines; the last to do so was the British and Foreign Bible Society, which brought out over 500,000 *Gleanings for the Young* in 1869.[86] In its first year of publishing children's periodicals, 1846, the Baptist Missionary Society sold 45,000 copies of its *Juvenile Missionary Herald* each month.[87] The London Missionary Society, one of the earliest in the children's publishing field, produced over 1,000,000 juvenile magazines a year by 1850. Sales at very cheap prices more than paid for publishing costs, which ran to over £1,350 annually by mid-century.[88] The Church Missionary Society published numerous children's periodicals, among them *The Church Missionary Juvenile Instructor*, begun in 1842, which catered to educated children and sold for one halfpenny, and *A Quarterly Token*, started in 1856, which went free to Sunday school subscribers of a farthing a week. By the end of the century the Society distributed about 900,000 copies of *A Quarterly Token* each year in England alone, and sold over 700,000 copies of *The Children's World*, the former *Church Missionary Juvenile Instructor*.[89]

Despite the particular sectarian interests of the children's missionary magazines, their contents were roughly the same: stories, reviews, poetry, hymns, natural science, and news of associations, Sunday schools, and foreign missions. All were sentimental in tone. And all of them stressed, over and over again, the themes of thrift and self-denial, the power of sin, the weakness of the flesh, salvation by faith, the

sionary associations, London, 1854; M. A. S. Barber, *Missionary Tales, for little listeners*, London, 1840.

[86] *The Sixty-Sixth Report of the British and Foreign Bible*, appendix, p. 119.

[87] *The Juvenile Missionary Herald*, i, 1845, preface.

[88] *The Report of the Directors to the Fifty-Sixth General Meeting of the Missionary Society*, London, 1850, lxxxix.

[89] Stock, *The History of the Church Missionary Society*, ii, p. 50; *Proceedings of the Church Missionary Society for Africa and the East. One-Hundred-and-First Year, 1899–1900*, p. 412. For information on other children's publications of the Church Missionary Society, including *Letters to Schoolboys, A Paper for Schoolboys, Seed-Time*, and *The Sower's Seed Basket*, see the Publications Sub-committee minutes, 1881–1914, Church Missionary Society Archives. I have included a few more nineteenth-century children's periodicals in the bibliography, most of which can be found in the British Library.

sweetness of a Christian death, the wickedness of the heathen, and the moral and cultural superiority of the British. Xenophobia was implicit, as was anti-Catholicism, which sometimes surfaced in the form of little moral tales in which the good Protestant triumphed over the hirelings of perfidious Rome.[90] The editors, many of them women by the late nineteenth century,[91] showed an almost morbid interest in Africa and Asia. Here compassion, not untouched by bigotry and gloom, pervaded such edifying tales as 'Six Deserted Chinese Babies',[92] 'Suttee',[93] 'Kaffir Cruelty',[94] 'Devil Worship',[95] 'The Pitiable Ignorance and Idolatry of the Heathen',[96] 'Savages of the South Seas',[97] and 'The Cannibal-Colporteur Gone Home'.[98] Such pieces must have put children in the proper frame of mind to appreciate the tales of horror recounted by missionaries and agents of the societies at the missionary meetings. The hymns that embellished the magazines and meetings were also instructive. Up to 6,000 children sang Mrs Gilbert's 'Hymn' at the Exeter Hall gathering in 1842:

Lord! while the little heathens bend,
And call some wooden god their friend;
Or stand and see, with bitter cries,
Their mothers burnt before their eyes;

While many a dear and tender child
Is thrown to bears and tigers wild;
Or left upon the river's brink,
To suffer more than heart can think;

[90] See, for example, 'No Popery' in *The Wesleyan Juvenile Offering*, viii, January 1851, pp. 11–12.
[91] Miss M. A. S. Barber, for example, edited the *Children's Missionary Magazine*, and Emily S. Elliott edited *The Church Missionary Juvenile Instructor.*
[92] *A Quarterly Token for Juvenile Subscribers*, October 1869, pp. 4–6.
[93] Ibid., January 1866, pp. 6–8.
[94] *The Wesleyan Juvenile Offering*, xiv, October 1857, p. 113.
[95] Ibid., v, n.s., 1871, pp. 44–7.
[96] *The Church Missionary Juvenile Instructor*, v, May 1846, p. 141.
[97] *The Juvenile Missionary Herald*, May 1859, pp. 71–2.
[98] *Gleanings for the Young*, i, n.s., January 1878, p. 2.

Behold, what mercies we possess!
How far beyond our thankfulness!
By happy thousands here we stand,
To serve thee in a Christian land.[99]

By such means, authoritarian moralists brought their children into contact with evil. The fixation with death and human corruption common in nineteenth-century popular literature, and not least in the missionary periodicals, moulded as it frightened children. 'There is no mission in Hell', warned *The Bible Class Magazine.*[100] Innocence, it seems, was to be protected by touching the frontiers of fallen, satanic life.[101]

The power of the missionary societies to form racial and cultural attitudes in the young should not be underestimated. In Britain this is perhaps their most enduring legacy. The scope of their activities was vast, for the massive number of publications and public addresses reached millions of impressionable children in the nineteenth century, through well-organized channels. Before the advent of state education in England, Sunday schools and missionary associations were among the most important agencies of indoctrination outside the home. If one could ask children active in the missionary movement at the end of the century what words they associated with 'African', for example, they would probably include: heathen, sin, black, slavery, fear, dark, ignorance, evil, and death. We must look more closely at the role played by children's literature and organizations in the formation of prejudice.[102]

Christian parents endorsed the work of the societies among the young, though not without reservations. Some of them believed that Sunday school funds could be better allocated.[103] And there was also worry lest poor children be

[99] *The Children's Missionary Meeting in Exeter Hall, on Easter Tuesday, 1842*, p. 14.

[100] *The Bible Class Magazine*, viii, 1855, p. 19.

[101] There is an interesting development of this theme in Peter Coveney, *The Image of Childhood*, London, 1967.

[102] The importance of Sunday school periodicals in the formation of racial prejudice is touched upon in Laqueur, *Religion and Respectability: Sunday Schools and Working Class Culture, 1780–1850*, p. 309. See also Douglas A. Lorimer, *Colour, Class and the Victorians*, Leicester, 1978, pp. 75–6.

[103] Blake, *The Day of Small Things*, pp. 34–5.

embarrassed by having to compete with better-off classmates for contributions.[104] On balance, however, parental criticism of the children's missionary movement was muted. Not surprisingly, for Sunday schools and juvenile associations, unlike state schools, were voluntary institutions over which parents had considerable control. Only so long as they kept children from corruption and educated them along lines acceptable to parents could they prosper. As we have seen, the charitable pursuits of children reflected those of their parents, especially their mothers, who, quite naturally, wanted their sons and daughters to be like themselves. The Reverend J. A. Page anticipated the day 'when the daughters of England shall become the Mothers of England, and bring their influence to bear both upon the family and domestic circle'.[105] At least one mother asked him to apprentice her daughters to the missionary cause, as he had apprenticed her many years before.[106] Through the missionary societies and, of course, other forms of charitable enterprise, such Christian mothers helped to ensure that the habits of authority and benevolence passed, without much question, from generation to generation. This was not the least of the legacies that benevolent women, guardians of the family and apostles among the young, left to the nation.

[104] In his earliest attempt to set up a juvenile society, Joseph Blake ran into difficulties over the inability of poor children to contribute. He was even accused of collecting their pennies for his own use. See 'Beginnings of Juvenile Missionary Associations', Notes and Transcripts, 6, Methodist Missionary Society Archive.

[105] *Monthly Reporter of the British and Foreign Bible Society*, ix, September 1, 1869, p. 51.

[106] *Seventy-Fourth Report of the British and Foreign Bible Society*, London, 1878, p. 204.

PART TWO:

The Power of the Cross

IV

In the Homes of the Poor

If philanthropy was widely held to be the most reliable remedy for the various maladies of English life, how was it to minister best to a population growing so rapidly that it threatened to overwhelm the means of subsistence, a population unsettled by industrialization and crowding willy-nilly into the cities and towns that blackened the English landscape? What innovations in philanthropy, particularly in its organization, could be devised to combat distress and the resulting social tension? Could the Church offer a solution? Could social science? Hospitals, dispensaries, reformatories, refuges, orphanages, and asylums had existed for a long time, but they had shown themselves incapable of checking the prodigious problems. Perhaps the problems could not be checked by such specialized remedial institutions. In the late eighteenth century new and various charities emerged which were to gain enormous public support in the nineteenth century—these were the visiting societies, and their aim was nothing less than to prevent distress and to promote social harmony.[1]

The custom of visiting the poor had very old traditions in England and in the western world generally. Scripture urged the practice and Christians, from Bishops to working women, carried out the biblical prescription. Though the nineteenth century was the heyday of the visiting society, the tradition of casual visiting remained very powerful, both in town and countryside. Agents of the societies often had

[1] Very little has been written on the visiting movement. For an introduction to the subject see David Owen, *English Philanthropy*, pp. 138–43; A. F. Young and E. T. Ashton, *British Social Work in the Nineteenth Century*, London, 1956, pp. 88–90; Walton, *Women in Social Work*, pp. 14–15, 57–8; and Anne Summers, 'A Home from Home—Women's Philanthropic Work in the Nineteenth Century', *Fit Work for Women*, ed. Sandra Burman, London, 1979, pp. 33–63.

experience as casual visitors and many continued this work. What the charities added to customary practice was system. Geared to cities and towns, where the suffering was most concentrated, they divided communities into districts, more often than not based on parish boundaries. Dividing the districts into streets, and the streets into households, they assigned visitors to each district, ideally one visitor to every twenty to forty families. Armed with the paraphernalia of their calling—Bibles, tracts, blankets, food and coal tickets, and love—these foot-soldiers of the charitable army went from door to door to combat the evils of poverty, disease, and irreligion. In other words, they sought to reform family life through a moral and physical cleansing of the nation's homes. In its thoroughness it was a system that must have warmed the heart of Jeremy Bentham.[2]

Visiting societies were well under way in England by the end of the eighteenth century. John Wesley's experience as a sick visitor in the 1740s resulted in much regular visiting by members of his fellowship;[3] full-blown societies, independent of church control, emerged somewhat later. It is difficult to determine which organization was first in the field, for many societies that may have practised visiting, particularly at the parish level, have left little or no record of their activities.[4] But among the earliest was the United Society for Visiting and Relieving the Sick, sometimes called the

[2] The Utilitarians adapted the system of district visiting to their own purposes. In 1813, the West Lancastrian Association passed a series of resolutions which included one proposing that a group of gentlemen volunteers make a house-to-house survey of the numbers of uneducated children in the City of Westminster. Edward Wakefield was largely responsible for the administration of the scheme, which included advice to parents on keeping their children clean. *The Philanthropist*, iii, 1813, pp. 361–3; *Parliamentary Papers* (hereafter cited P.P.) *Reports from Committees, Report from Select Committee on the Education of the Lower Orders of the Metropolis*, 1816, iv, pp. 40–2.

[3] In a sermon 'On Visiting the Sick', Wesley had called on Christians, rich and poor alike, to give whatever time they could spare, 'for the relief and comfort of their afflicted fellow-sufferers'. In these forays into the homes of the poor the principle unit of organization was the class-meeting; the principle enemies, in accord with Wesley's views on the matter, were sloth, dirt, and tea. *The Works of The Rev. John Wesley*, vii, p. 123; Eric McCoy North, *Early Methodist Philanthropy*, New York, 1914, pp. 36–40.

[4] There were societies in the mid-eighteenth century which distributed tracts from door to door, for example the Society for Promoting Religious Knowledge, established in 1750.

Willow-Wick Society, founded in Moorfields in 1777. This virtually forgotten charity, whose religious affiliation is uncertain, raised more than £2,000 in its first seventeen years and contacted 16,000 cases.[5] Other charities entered the field in the early 1780s. In London there was the Friendly Society or Charitable Fund for the Relief of the Sick Poor at their own Habitations (1781)[6] and in Norwich the Society of Universal Good Will (1784).[7]

The beginning of district visiting on a significant scale is probably best dated from 1785, when a Methodist, John Gardner, established the Benevolent, or Strangers' Friend Society in London.[8] In its rules and organization this influential charity became a model for many of the visiting societies that followed, both Methodist and of other denominations. As with so many philanthropists it was Gardner's firsthand experience of death and hardship that jolted him into action: 'I had been visiting a poor man dying of fistula. He lay on the floor covered with a sack, without shirt, cap, or sheet, and in a dull, despairing tone exclaimed, "I must die without hope." '[9] Gardner related his experience to his wife and they induced several of their neighbours to join them in forming a visiting charity. Supported by Wesley it grew rapidly, and by 1802 it boasted a distinguished committee (all male) and 110 visitors working throughout the metropolis. From an income of over £1,000 it relieved nearly 2,000 families in that year.[10] Like the many other Methodist Strangers' Friend Societies that were established in the early nineteenth century, Gardner's sought 'to relieve the destitute sick poor, without distinction of sect or country'. It focused its attention on those not entitled to parochial relief, especially workers paid on a weekly basis, widows left without families, and, to use

[5] *The Wesleyan Methodist Magazine*, i, 1845, pp. 661–2.

[6] North, *Early Methodist Philanthropy*, p. 47.

[7] B. Kirkman Gray, *A History of English Philanthropy*, London, 1905, p. 169; *Articles and Regulations proposed for the Society of Universal Good-Will, in London, or Elsewhere, 1789*, Norwich, 1789. See also Owen, *English Philanthropy*, p. 67.

[8] Sampson Low, *The Charities of London*, London, 1850, p. 117, cited the Strangers' Friend Society of London as the oldest institution of its kind in England.

[9] Quoted in North, *Early Methodist Philanthropy*, pp. 47–8.

[10] *The Nature, Design, and Rules of the Benevolent or Strangers' Friend Society*, London, 1803, pp. 6, 7, 47.

the Society's words, those 'of a more profligate and vicious cast'.[11]

By 1815 there were a variety of other visiting charities working alongside those established by the Methodists. Some were widely publicized, like the Widow's Friend and Benevolent Society (1808) and the Spitalfields' Benevolent Society (1811).[12] Others were less conspicuous, like the Misericordes Society (1812), which sought to relieve the sick poor of St. Swithin, London Stone.[13] More obscure, but no less interesting, was a charity founded in Seven Dials sometime in the first decade of the nineteenth century. It called itself the West Street Chapel Benevolent Society and was set up by the poor to relieve their poor neighbours. Details of its operations are few, but it appears to have been influenced by Sir Thomas Bernard's innovative Society for Bettering the Condition of the Poor, which had built the West Street Chapel in 1800. A committee of twelve poor persons met weekly with their visitors to inspect the books and proffer relief. Their aims were straightforward: 'to instruct the ignorant, to exhort the careless, to comfort the feeble-minded, and to glorify God in all things', which in practice boiled down to Bible reading, advice on cleanliness, and medical attention provided by a local dispensary.[14] Unfortunately, it will probably never be known just how many societies there were like this one, the formal expression of working-class visiting traditions. But we should not underestimate their number. In the 1850s a working men's pastoral aid society in Liverpool, assisted by wealthy citizens, could boast 2,000 subscribers.[15]

As far as one can tell, committees made up entirely of men managed all the above-mentioned charities, though the visitors were often women. So when did the first visiting

[11] Ibid., p. 6. There were 'Strangers' Friend Societies' in most parts of the country, including Bristol, Birmingham, Manchester, Liverpool, Sheffield, Leeds, York, and Hull. *The Wesleyan Methodist Magazine*, i, 1845, pp. 661–8.

[12] See, for example, *The Times* during the winter of 1820–1.

[13] Anthony Highmore, *Philanthropia Metropolitana*, London, 1822, pp. 330–2.

[14] Anthony Highmore, *Pietas Londinensis*, 2 vols., London, 1810, ii, pp. 920–1.

[15] A. Hume, *Analysis of the Subscribers to the Various Liverpool Charities*, Liverpool, 1855, p. 11.

societies run by women appear? There is a tantalizing reference to a 'Ladies Charity' in Soho, founded in 1773, to relieve sickness and furnish employment, but whether or not it practised visiting cannot be determined.[16] One of the earliest female societies which has left a reliable reference to visiting appeared in 1791: the Lying-in Charity of Tottenham. This small parish charity initially limited its services to supplying midwives and linen to women in their confinement, but soon extended its work to the sick poor, providing linen, money, food, coals, and medicines. It relieved 14 women in its first year, a figure which rose to 94 in the seventh year.[17] These were very small figures by the standards which were to follow, yet the ladies of Tottenham, perhaps for the first time, had organized a traditional activity of women, sick visiting, on a methodical basis. Their society gave timely assistance to the sick poor in the parish; at the same time it encouraged more women of superior rank to become acquainted with the habitual distresses of their less fortunate neighbours. By the early years of the nineteenth century similar female societies had set up operations in cities and towns around England.[18]

Most of the early female visiting societies specialized in lying-in and sick visiting. But several of them branched out into more general visiting in the early nineteenth century. In 1802 the Friendly Female Society, patronized by the Princess of Wales, was formed in London to relieve infirm and aged women on small incomes.[19] In 1810 the Ladies' Benevolent Society of Liverpool began to supply food and clothing to relieve that city's needy poor.[20] And two years

[16] Highmore, *Pietas Londinensis*, ii, p. 944.

[17] Ibid., i, pp. 389–92. For a further discussion of this charity see The *Reports of the Society for Bettering the Condition and Increasing the Comforts of the Poor*, ii, pp. 26–9. The ladies of Tottenham were also pioneers in founding savings banks for the poor. See *Annals of Banks for Savings*, London, 1818, pp. 11, 51–2.

[18] Including Cambridge, Ipswich, Colchester, Needham, Halstead, Stoke Newington, Edmonton, Cheshunt, Ware, Hertford, Woodford, Wanstead, Walthamstow, Canterbury and Liverpool. Highmore, *Pietas Londinensis*, i, pp. 389–92 and *passim.*

[19] *The Friendly Female Society, for the Relief of Poor, Infirm, Aged Widows, and Single Women*, London, 1803.

[20] *The Fifth Annual Report of the Ladies' Benevolent Society, Liverpool*, Liverpool, 1815.

later the fashionable Ladies' Royal Benevolent Society began its investigations into the condition of the London poor.[21] The final seal of approval to the efforts of women to organize their customary visiting came when Queen Charlotte agreed to be its patroness. By the end of the Napoleonic wars there were scores of female visiting societies, most of them small parish charities. Along with the societies managed by men, in which women acted as visitors, they played an increasingly prominent part in the nation's health and well-being. More and more communities welcomed them as a reliable antidote to the problems of poverty, disease, and childbirth; they proved especially useful in rapidly expanding cities short of medical facilities. At their best they were thorough, cheap, and preventative. And like the casual visiting that persisted into the industrial age, they bolstered that network of relations between the classes which, for better or for worse, gave England a semblance of social order in the early years of the nineteenth century and after.

It is fashionable to suggest that fear of 'Gallic liberty' and domestic unrest gave a great stimulus to the growth of such societies in the 'counter-revolutionary decades'. Visiting was, of course, designed to put the establishment in closer touch with the working classes; and Sarah Trimmer, for one, heartily recommended the women of England to do a house-by-house inspection of poor neighbourhoods as a means of averting social unrest.[22] Yet this motive can be overstated. Many visiting charities were organized before the outbreak of the French Revolution and remained active long after any threat existed from the continent, or for that matter long after any serious threat of working-class agitation existed at home. As one historian has argued, fear of revolution may have been behind the great increase in charitable subscribers, 'but the persistance of their contributions through changing circumstances shows that their ruling motives were genuine philanthropy and a steadfast determination to perform what they believed to be their

[21] *The Ladies' Royal Benevolent Society*, London, 1818.

[22] Sarah Trimmer, *The Œconomy of Charity*, 2 vols., London, 1801, ii, pp. 57–9.

duty'.[23] The French Revolution frightened most England philanthropists to be sure, but it did not usually unbalance them. The visiting societies of the late eighteenth and early nineteenth centuries did not wish to excite political prejudices among the poor and so advised their visitors to avoid political discussions. In any case visitors had enough on their hands in dealing with such perennial problems as lying-in, the sick, the widowed, and the aged.

In time there were as many types of visiting society as there were denominations and distresses. The evangelicals set the tone, and as one of their number reminds us, they were on the alert to 'check . . . the insidious approaches of our Jesuitical foe'.[24] Sectarian rivalry was intense among the visiting charities, for Roman Catholics, Tractarians, Jews, Unitarians, and even non-denominational visitors were active. The deaconesses, whether High Church or crypto-Catholic, did a great deal of visiting from their houses of mercy in the slums. One band of French Catholic peasant women, 'the little sisters', begged and lived off scraps themselves to visit the poor in Hammersmith.[25] On the other side of London but in a different world altogether, Lady Battersea, born a Rothschild, established the Jewish Ladies' Benevolent Loan Society, which visited East End Jews. The claims on this Society's services were slight, if we are to believe one of the Jewish mothers visited by the charity: 'We are not like the *goyem* (Gentiles), we do not want to be talked to or taught, we do not drink, and we know how to bring up our children religiously and soberly.'[26]

The largest societies canvassed entire cities and looked into a wide range of problems. Others assisted only distressed widows, or needlewomen, or the aged. There were forty-five visiting charities to the blind alone by 1889, which were the pioneers in social work among the blind.[27] Some of the visiting societies described themselves by what they provided:

[23] E. C. P. Lascelles, 'Charity', *Early Victorian England, 1830–1865*, ed. G. M. Young, 2 vols., London, 1934, ii, p. 347.

[24] *The Harbinger*, September, 1853, p. 283.

[25] Margaret Goodman, *Sisterhoods in the Church of England*, London, 1863, pp. 236–42.

[26] Constance Battersea, *Reminiscences*, London, 1922, pp. 414–17.

[27] Young and Ashton, *British Social Work in the Nineteenth Century*, p. 188.

coals, food, or clothing. At least one charity distributed quilts inscribed with biblical quotations. What the working man felt who awoke to see the seventh commandment on his 'silent comforter' is a matter for conjecture.[28] There were societies connected to homes, missions, hospitals, dispensaries, and temperance societies.[29] The largest single type was connected to the parish. Some of them came and went with the seasons, being founded in the winter to combat famine or fever only to be dissolved in the spring when the weather changed and the condition of the poor improved. But most societies, whether parish-based or larger, persisted even in the best of times. What they all had in common was the system that formed the basis of their work, the visitation of the poor in their own homes.

In the relative calm of the mid-Victorian years there were hundreds of visiting societies in metropolitan London alone. Some of them were large and richly endowed, like the General Society for Promoting District Visiting and the Strangers' Friend Society. The London City Mission carried out about 2,000,000 visits a year on a budget of between £20,000 and £40,000 annually.[30] At the other extreme there were miniscule societies, like the Aged Couples' Charity, which could boast only two visitors with very little money to look after the infirm poor of St. Mary's parish, Marylebone.[31] Virtually every London parish church sponsored a visiting society. When Sampson Low Jr. compiled his figures in 1850, he counted only thirty-six parish churches which did not; most of these were in rich neighbourhoods where charitable workers were in little demand.[32] There were over 250 parish churches in London by this time, plus nearly one hundred episcopal chapels,

[28] [Ellen Ranyard], *Nurses for the Needy or Bible-Women in the Homes of the London Poor*, London, 1875, pp. 224-5.

[29] For an example of district visitors associated with a temperance charity see [Mrs J. B. Wightman], *Annals of the Rescued*, London, 1861, pp. 213, 258.

[30] *The Thirty-Fifth Annual Report of the London City Mission*, London, 1870, pp. vii–viii. For a colourful account of the work of this society see R. W. Vanderkiske, *Notes and Narratives of a Six Years' Mission, principally among the dens of London*, London, 1852.

[31] W. H. Fremantle, *St. Mary's, Bryanston Square: Pastoral address and report of the charities of the year 1870*, London [1871], p. 41.

[32] Low, *Charities of London*, pp. 127–8.

which also commonly supported visiting charities. In the last quarter of the century, the Metropolitan Visiting and Relief Association, which promoted Church of England visiting societies by grants to London clergy, oversaw the activities of over two hundred of these parish charities with 2,200 visitors.[33]

In the second half of the nineteenth century the Catholic Church was no less committed in London. It supported scores of visiting societies and religious communities that visited. Sisters of Charity, Sisters of Mercy, and Sisters of Nazareth were active in various parts of the city; and complementing their work was an association called the 'Ladies of Charity', made up of women serving their local Catholic churches in much the same capacity as the women who formed parish visiting charities in the Church of England. In the Archdiocese of Westminster, for example, the Ladies of Charity could boast twenty visiting societies, most of them founded in the 1840s and 1850s.[34] It is uncertain how many of the roughly three hundred dissenting chapels in London by 1850 supported visiting societies as well, but it is quite likely that a majority of them did.

For all this diversity there was considerable uniformity in the constitutions which governed the various societies. The great majority of them supported themselves from voluntary contributions, largely subscriptions, and were managed by a committee elected annually by the members, who were normally required to contribute a guinea a year. The larger charities could boast an array of officers, including a president, vice-president, secretaries, treasurers, perhaps a surgeon or a nurse, and, at the bottom of the hierarchy, a host of visitors. Typically, the committee outlined and organized the districts to be canvassed. If it were a society managed by men, a ladies' sub-committee was usually formed to look after what was described as female work.

[33] J. C. Pringle, *Social Work of the London Churches*, London, 1937, pp. 190–1.

[34] *Handbook of Catholic Charitable and Social Works*, London, 1905, pp. 12–14. There are circulars announcing the setting-up of various Catholic ladies' visiting associations in London at Archbishop's House Archive, Westminster, File W.1/2.

The larger societies often appointed district sub-committees to oversee the administration in the particular neighbourhoods. This division of authority, it was argued, had the advantage of keeping the management in touch with local conditions. The sub-committees, which had their own secretary and treasurer, appointed and supervised the visitors, apportioned relief, and settled accounts. They normally held weekly meetings and in turn reported to the general committee each month. At the end of the year, or at the time of the annual general meeting, the societies that could afford it published their reports, which included objectives, rules, funds received and expended, cases relieved, lists of subscribers, and perhaps a few case-studies.

Despite the similarities in organization, diversity and rivalry remained hallmarks of the visiting societies. Institutions like the Metropolitan Visiting and Relief Association (1843), the Society for the Relief of Distress (1860), and the Charity Organisation Society (1869) sought to bring visiting charities into closer contact and offered them assistance in using their resources more efficiently. But because of sectarian squabbles and a rapidly growing and shifting population, not to mention the usual human failings, the challenge of rationalizing charitable resources and methods was never fully met. Philanthropic enterprise was, in a sense, *laissez-faire* capitalism turned in on itself. There were few restrictions placed on the charitable contributor, and in a society splintered by class, local, and religious allegiances, charities proliferated. Curiously enough, they competed for the custom of the poor.

The overlapping of rival charitable authorities was staggering in some cities. In parts of London, for example, High- and Low-Church Anglican, Catholic, Methodist, Nonconformist, and non-denominational visitors were working in the same field. With thousands of visitors entering hundreds of thousands of households each year in London, few poor families were free from their dutiful attentions. Indeed, a family might be visited several times each month, by a host of visitors, or intruders, depending on their perceptions. Some visitors, taking advantage of their superior social status, knocked and entered before the door could be opened.[35] If the visitors from the great societies were not enough to bemuse, pester, or redeem the poor, there were those of the smaller parish societies. Then there were casual visitors, perhaps the vicar's wife, or a better-off neighbour, or the seasonal attentions of rich ladies with little on their minds who found 'slumming' a distraction from their privileged routine.[36] Was this bewildering array of charitable activity not tantamount to an invasion of an Englishman's privacy, asked one critic?[37]

To complicate matters further, the visiting societies also competed with the Poor-Law authorities. The New Poor Law of 1834 appears not to have had any marked effect on the growth of subscriptions to the visiting charities, though as a 'charter of the ratepayer', to use G. M. Young's phrase, it may have stimulated contributions. Certainly it altered the relationship between poor relief and the voluntary societies. A guiding principle of the New Poor Law was the abolition of outdoor relief to the able-bodied; and as much of the assistance provided by district visitors before 1834 supplemented either wages or outdoor relief, confusion and conflict were unavoidable. Many visitors were oblivious to the new Act and contrary to its principles continued to subsidize the poor with cash payments. Others knew its character only too well and consciously sought to subvert it. Keeping their charges out of the 'bastille' was highly desirable from the visitors' point of view, not least because they had less power over them once they were inside.[38] This was felt particularly by women visitors, who knew from bitter experience the difficulty they had in getting into workhouse wards. Thus visitors often worked at cross purposes with Union officials and Guardians, which added to the chaos in poor relief. As time passed, however, more and more visiting societies sought

[35] Claudia [Mary Pryor Hack], *Consecrated Women*, London, 1880, p. 273.

[36] See, for example, *Reminiscences; by a Clergyman's Wife* [Fanny Alford], ed. The Dean of Canterbury, London, 1860, Chapter 1.

[37] Anon., *District Visitors, Deaconesses, and a proposed adaptation, in part, of the Third Order*, London, 1890, p. 7.

[38] See Summers, 'A Home from Home—Women's Philanthropic Work in the Nineteenth Century', *Fit Work for Women*, pp. 46–7.

to reconcile their differences with the Poor-Law authorities and took measures to discipline their visitors.

The societies were only as good as their visitors, and so they made every effort to enlist moral, zealous, and intelligent recruits. But they could not often afford to be choosy, for visitors were normally part-time volunteers. They usually came from the neighbourhoods they patronized and were likely to be associated with the local church or chapel, or a charity school or mothers' meeting. Visiting charities had a pronounced local character. As we have seen, most of them were based on the parish; but even the largest societies depended heavily on their neighbourhood branches. This suited nineteenth-century English society very well, for it was a relatively unadministered world, of 'calling' on neighbours and 'leaving cards', in which the pressures on individuals were typically local and immediate. Women felt the pressure to contribute to local causes especially, for their lives were commonly wrapped up in their immediate surroundings and given over to serving family and friends. In the face-to-face world of the community the inducement to 'keep up with the Joneses' could not be taken lightly. The visiting societies counted upon it for their recruitment.

The problem of recruitment was more serious in those areas where few people had the leisure to volunteer. An 'immense geographical gulf' existed between the classes in many English cities. Middle-class residents were alarmingly underrepresented in East London, for example (the Settlement movement sought to correct this in the 1880s). This separation of the classes, it has been argued, demoralized the poor and 'upset "the balance" between charity and the Poor Law'.[39] It also created problems for the visiting societies. Some of the neighbourhoods had the reputation for being too dangerous for respectable women to enter. Complaints about the squalor, the danger of infectious disease, and the travel time to and from a district were common; and, it appears, some women worried about the number of working men 'lounging about the smaller

[39] Gareth Stedman Jones, *Outcast London*, Oxford, 1971, pp. 247–51.

streets'.[40] Because of such objections many women visitors, in London for example, preferred the settled calm of the West End or Highgate to the seediness of the East End or Seven Dials, where they were needed most. Yet well-to-do women, sometimes recommended to go in pairs, braved the dens and rookeries of the most run-down districts. Quite a few societies, among them the deaconess institutions, enlisted working women to staff their visiting ranks.[41] In a few instances middle-class women took it upon themselves to hire working men to canvass poor neighbourhoods.[42] The charitable organizers in poor urban parishes would use anyone willing to work, but found it easier to get humble women than humble men.[43] All in all, visiting attracted and brought together a wide range of abilities and classes.

What proportion of visitors were female? Virtually all were in visiting societies managed by women, deaconess institutions, and Catholic sisterhoods. It is more difficult to establish the precise percentage in those charities managed by men. The London City Mission was exceptional in using paid male 'missionaries'.[44] (If regular male visitors were required it was probably necessary to pay them.) The internal evidence and the remarks of authors of district-visiting handbooks suggest that the ratio of female visitors to male was quite high in most societies, perhaps two or three to one by the mid-century, and growing higher all the time. Charles Bosanquet, the first Secretary of the Charity Organisation Society, used the masculine pronoun when writing about

[40] Charles B. P. Bosanquet, *London: Some Account of its Growth, Charitable Agencies, and Wants*, London, 1868, p. 129; Dame Katharine Furse, *Hearts and Pomegranates. The Story of Forty-Five Years, 1875–1920*, London, 1940, p. 156.

[41] Anon., *The House of Mercy at Ditchingham*, Oxford and London, 1859, pp. 16–17; Carter, *Is it well to institute Sisterhoods in the Church of England?*, p. 5.

[42] [Wightman], *Annals of the Rescued*, p. 239; The Revd. Buckley Yates, *Miss Shepherd, of Cheadle, Staffordshire*, Manchester, 1876, p. 90.

[43] Gregory, *The Difficulties and the Organization of a Poor Metropolitan Parish*, p. 32. For an example of the work of humble women visitors in the East End of London see the Revd. Frederick W. Briggs, *Chequer Alley*, London, 1866.

[44] In 1835 the missionaries were paid on average £60 a year. *First Report of the London City Mission*, London, 1835, p. 7. The wives of missionaries were expected to help in the work of the society without pay. See *These Fifty Years. Being the Jubilee Volume of the London City Mission*, London [1884], pp. 136–44.

visitors in 1874, but he was acutely aware that the 'great majority' of them were women.[45] 'Men', he lamented some years earlier, 'and gentlemen especially, are doing very little in proportion to their numbers'.[46] When Charles Loch, Bosanquet's successor at the COS, wrote a guide to visitors in 1882, he dropped the masculine pronoun altogether and used 'she' when referring to visitors.[47]

Despite the prejudice against them in some quarters, women showed, from the beginning of the century, a greater willingness to work for charity than men and an ability to handle a wide range of cases. Some societies, the Bedford Institute for example, admitted that visiting was best done by women, 'as they can freely enter the homes in which few men are ever found, except when confined by illness'.[48] To use the language of Mary Bayly, a leading charitable campaigner, respectable wives and mothers were ideally suited to mend the nation's ragged homes.[49] This rationale was compelling. One of the secrets to successful visiting was a knowledge of domestic management; everyone agreed that female volunteers moved more easily amongst wives and mothers and were more sympathetic to their problems; and some argued that they were also more likely to uncover female dissimulation. Moreover, the use of women as visitors was in accord with society's deeply ingrained beliefs about the family. The protection of the family was the cornerstone of nineteenth-century social policy; within the family the role of wife and mother was thought crucial. 'Reform the mothers and you begin at the root of our social evils', remarked a report from a female visiting charity in Westminster.[50] Few would have disagreed. It was a maxim certain

[45] Charles B. P. Bosanquet, *A Handy-Book for Visitors of the Poor in London*, London, 1874, p. vii.

[46] Bosanquet, *London: Some Account of its Growth, Charitable Agencies, and Wants*, p. 111.

[47] C. S. Loch, *How to Help Cases of Distress*, London, 1883, pp. 16–17.

[48] *Bedford Institute, First-Day School and Home Mission Association. Report*, London, 1867, p. 14.

[49] See Mrs [Mary] Bayly, *Ragged Homes, and how to mend them*, London, 1860.

[50] *Report of the Visitation of Females at their own Homes, in the City of Westminster*, London, 1854, p. 4.

on the same table; another child was dangerously ill; the poor man disabled by severe rheumatism; and his wife in a state of distraction, occasioned by a long series of afflictions and painful privations. The relief afforded came very opportunely; and the Visitor has had the pleasure of seeing the father and sick child restored to health, and the mother to a comparative state of composure of mind. It is not too much to say, that they have been raised by the Society from a state bordering on despair, and from the brink of perishing. [Spitalfields, 1820] [65]

My then minister requested me to accompany him on a visit to a case of particular interest,—that of a young girl of seventeen, blind, deaf, and dumb; often suffering much pain, and, with the exception of her right hand, entirely paralysed . . . the only method of communicating with her is to pass the fingers of her right hand over the letters of a raised Alphabet, and so spell out word after word. [Brighton, 1858] [66]

In visiting a slum in a town in the North of England, our Officers entered a hole, unfit to be called a human habitation—more like the den of some wild animal—almost the only furniture of which was a filthy iron bedstead, a wooden box to serve for table and chair, while an old tin did duty as a dustbin.

The inhabitant of this wretched den was a poor woman, who fled into the darkest corner of the place as our Officer entered. This poor wretch was the victim of a brutal man, who never allowed her to venture outside the door, keeping her alive by the scantiest allowance of food. Her only clothing consisted of a sack tied round her body. Her feet were bare, her hair matted and foul, presenting on the whole such an object as one could scarcely imagine living in a civilised country . . . .

We took the poor creature away, washed and clothed her; and, changed in heart and life, she is one more added to the number of those who rise up to bless the Salvation Army workers. [late nineteenth century] [67]

There were countless cases, equally distressing, throughout the nineteenth century, and it must be appreciated that if they were not attended to by charity, many of them would not have been attended to at all. The poverty was so profound, the disease and hardship so overwhelming, that abstract debates about the value of charity were often out of

[65] This case was cited in an advertisement from the Spitalfields' Benevolent Society, *The Times*, Dec. 28, 1820.

[66] Sarah Robinson, *Light in Darkness*, London, 1859, pp. 5–6. The fate of this girl is discussed in [Sarah Robinson], *The Darkness Past*, London, 1861.

[67] General William Booth, *In Darkest England and the Way Out*, London, 1890, p. 191.

place. The visitors who came across such scenes did not have the time, indeed the detachment, to question the nature and the ultimate result of the relief they provided. Nor could they make distinctions between the deserving and the undeserving in such cases. To some women visitors such distinctions were a nonsense in the name of political economy.[68] When confronted with death and dying, they acted in the only way they knew, with compassion and whatever relief they could bring to bear, and moved on.

The visited, especially those who received medical assistance or the necessities of life from visitors, were often touched. Memoirs and charitable reports are full of their letters. 'I can't tell you . . . what that lady has done for me', wrote a miner about Mary Shepherd, a Methodist visitor in Cheadle, 'I owe my life to her. If it had not been for her I should have been in my grave. When I was crushed in the mine by a mass of earth falling upon me, and all the bones seemed broken in my body, she supplied me with every necessary thing that I wanted. . . . I shall never forget her kindness to me.'[69] The wife of a factory operative wrote to the Shrewsbury teetotal visitor Mrs J. B. Wightman: 'I am thankful to God that ever I had knowledge of you; for, through being induced to join the pledge, I have been rescued from destruction.' Another labourer's wife wrote to Mrs Wightman: 'This is to certify that it is to you, under God, that I owe all my present comforts, and these are not a few in comparison of what I had two years ago, when first I became acquainted with you.'[70] At their best, visitors earned the blessing and regard of the grateful poor.

It was the seriousness of their task that made the visiting societies obsessed with feigned distress. It was an obsession all the more disturbing because charity itself bred deception. Given the number of visitors in the field, it is not surprising that the poor sometimes took advantage of them. Visitors

[68] See for example, An Old Maid [Lucy F. March Phillipps], *My Life and what shall I do with it?*, London, 1860, p. 239.

[69] Yates, *Miss Shepherd, of Cheadle*, p. 111.

[70] [Wightman], *Annals of the Rescued*, pp. 242–4. Ellen Barlee, *A visit to Lancashire in December, 1862*, London, 1863, pp. 152–6, includes a series of letters from operatives to a clergyman, some of which show their appreciation for past services, especially night-school classes.

often encouraged this by their casualness and thoughtlessness. But we must remember something which those who worked for nineteenth-century charities never forgot: there was not enough money to go around, and that those in need were often so desperate that funds must not be wasted on fraudulent cases. When it was common for from 5 to 10 per cent of those admitted to a visiting society's books to die soon after, impostors took on a sinister importance.[71] Swelling the category of the undeserving, they gave real meaning to the term 'deserving poor'. One writer estimated in 1838 that impostors wrote one thousand begging letters each day in London alone.[72] Other ingenious adventurers dressed up in ragged clothes and touching lies and visited the neighbourhoods of the rich.[73] But perhaps the most common type of impostor was the man or woman who could tell a fictitious tale of woe so movingly that it opened the visitor's purse. Women visitors, who were in some circles thought to be an especially easy prey to such dissimulation, were cautioned repeatedly. Only a careful analysis of each case could prevent such deception.

But what kind of woman took on the often harrowing routine of visiting? And what was it that led her to volunteer? The temptation here is to fall into caricatures, the 'evangelical girl of the period',[74] or the visiting types amusingly drawn by the author of *My District Visitors*,[75] or perhaps the literary portraits more familiar to us. There were 'morally tidy' and unbending visitors in the style of Mrs Pardiggle, Miss Gladwin, and Drusilla Clack, whose faith resembled their stockings, 'both ever spotless, and both ready to put on at a moment's notice'![76] But this type was

[71] See, for example, *Reports of the Society for Bettering the Condition and Increasing the Comforts of the Poor*, i, p. 232. Also [Ranyard], *Nurses for the Needy*, p. 250.

[72] James Grant, *Sketches in London*, London, 1838, p. 4.

[73] *The Evangelical Magazine*, xiv, 1806, p. 616. On a lighter note, Charles Dickens and his mates pretended to be poor boys and roamed the streets asking the passers-by, especially old ladies, for money. One lady told them she 'had no money for beggar boys'. John Forster, *The Life of Charles Dickens*, London [1879], p. 21.

[74] *The Girl of the Period Miscellany*, May 1869, pp. 92-4.

[75] See The Frontispiece.

[76] Wilkie Collins, *The Moonstone. A Romance*, 3 vols., London, 1868, ii, pp. 73-4.

probably less prevalent than critics like Dickens, Mary Bridgman, or Wilkie Collins would lead us to believe. To balance such portraits we must remember the softer more sympathetic benevolence of Agnes Gray, or old Alice in *Mary Barton*, or of the May girls in *The Daisy Chain*. We should not underestimate the influence of fictional visitors on nineteenth-century women; the 'constitutionally charitable' Lady Belfield in *Coelebs in Search of a Wife* made visiting the 'rage' according to Lucy Aikin.[77] But to understand the woman visitor we must look at the experiences of a few of them not touched up by writers of fiction.

Mrs Jane Gibson, a Methodist from Newcastle who visited for the Strangers' Friend Society in London, is typical. Her *Memoirs* are full of Scripture and citations from contemporary Christian authorities. Hannah More is quoted with enthusiasm: 'Charity is the calling of a lady; the care of the poor is her profession.' This calling flowed from 'that cardinal doctrine of the Gospel, . . . the vicarious suffering of Christ'.[78] The supreme charity of Jesus in sacrificing his life had the most poignant meaning for Bible Christians like Mrs Gibson, for it pardoned their sins and offered them salvation. Armed with passages of Scripture countless like-minded women went forth to tread in the steps of the 'Heavenly Visitor',[79] to express their gratitude for his unbounded goodness. The language of their hearts was 'what shall I render unto the Lord for all his benefits toward me'? (Psalms 116: 12.) Their most essential qualification was a love of Christ; and guided by his example, they opened the door of mercy to their fellow sinners. They loved the poor because Jesus loved the poor. This indirect nature of their love helps to explain its often abstract, impersonal quality. In the memoirs of women like Mrs Gibson love appears to have been manifested rather indifferently to everyone.[80]

[77] *Correspondence of William Ellery Channing, D. D. and Lucy Aikin*, ed. Anna Letitia le Breton, London, 1874, p. 396. Hannah More's *Coelebs* was an immensely popular and influential 'novel' full of evangelical doctrine dressed up in the guise of fiction.

[78] Francis A. West, *Memoirs of Mrs. Jane Gibson*, London, 1837, pp. 334–5.

[79] *Meliora*, vii, 1864, p. 242.

[80] See Beatrice Webb's description of her mother's lifelong companion Martha in *My Apprenticeship*, p. 21.

Christ had expiated mankind's sins, yet the reconciliation of man and God could only be accomplished by bringing men to 'the foot of the cross', bringing them, in other words, to a recognition of their own corruption. The death of their Redeemer showed Christians the value of their own souls, and thus led them to a greater diligence in securing their own eternal happiness. It also inspired them with a hatred of sin. From women's memoirs we get the impression that an evangelical life was a constant battle with temptation and the powers of darkness. The well-connected Anna Maria Clarke, like so many evangelicals (she was Anglican), went through a prolonged period of self-loathing and despair. She read Hannah More on the depravity of human nature. But it was Wilberforce's *Practical View* which brought her to Christ and awakened her interest in philanthropy. 'All our works are imperfect', she wrote in her diary at sixteen, 'but this faith must produce good works'.[81] On moving to Torquay in 1812, Mrs Clarke found an outlet for her spiritual anxiety: 'She longed to be made the instrument of awakening the "dead in trespasses and sins"; of seeking out the lost, of comforting the mourners; of strengthening the weak; of warning the backsliders; and of reproving the wicked.'[82] She became a visitor, distributed tracts, established a school for poor children, and set up a Bible society to supply the Navy. Philanthropy was not only a sign of her salvation; it helped to overcome her sense of sin. She would save her own soul as she saved the souls of others.

Women from very different backgrounds went through similar experiences. Mrs Margaret Burton, the daughter of an ironmonger, was born in Durham in 1785. On the death of her father four years later she and her mother moved to North Shields where they took up dressmaking. In 1802 Margaret gave herself to God at a Methodist class-meeting: 'Satan seems to have run through all his general temptations to hinder the soul panting for God.'[83] When she moved to

[81] The Revd. Thomas Grey Clarke, *A Memoir of Anna Maria Clarke*, London, 1853, pp. 17–18.

[82] Ibid., p. 68.

[83] John Dungett, *Life and Correspondence of the Late Mrs. Margaret Burton*, Darlington, 1832, pp. 27–8.

Darlington, she began a diary full of remorseful descriptions of her struggles with Satan. Her heart was like 'a cage of unclean birds'. And, in a much used metaphor, she asked 'how shall a guilty worm draw nigh to God'?[84] Conscious of her own sin she became conscious of the sin around her. To conquer it she turned to visiting:

> I furnished myself with some tracts which I distributed, and spoke to them as the Lord enabled me to do, though I am afraid it was not wholly without fear of giving offence; thence doing the work of the Lord deceitfully. Some received me kindly, one family shut their door against me, but I went again; and one poor old sinner was unkind; however I invited them all to come to my house every Sunday afternoon, and promised if spared to read to them. If the Lord hath himself set me to this work. I shall have His help and blessing.[85]

Like so many nineteenth-century women hers was a life of hard work and religious service, not all of it, as she related, appreciated. Yet she persevered, perhaps remembering Paul's advice to the Galatians: 'Let us not be weary in well-doing: for in due season we shall reap, if we faint not.' (Galatians 6: 9.) On her deathbed, at the age of forty-five, she passed on the duty of good works to her nineteen-year-old daughter Mary: 'Does my Mary cast all upon Christ?' she asked.[86]

We can answer her question, for a life of Mary exists. Its title, *Holy Living*, would have given Mrs Burton enormous relief. As we discover from it, Mrs Burton had insured her children against corruption; Mary was attentive at public worship by the age of two. Despite 'being sensibly alive to everything that was evil', she began backsliding at the age of eight.[87] Family prayer and class-meetings restored her to Christ and before long she was at boarding-school, where, we are told, she spent much of her time saving other pupils. Yet she had set-backs. Unable to extinguish the 'carnal mind', her *Devotional Remains* tell a story of guilt, 'heart-burning',

[84] Ibid., pp. 57–8. Another woman remarked 'He [God] deals very bountifully with a poor worm.' See Agnes Bulmer, *Memoirs of Mrs. Elizabeth Mortimer*, London, 1836, p. 193.

[85] Dungett, *Life and Correspondence of the Late Mrs. Margaret Burton*, pp. 73–4.

[86] Ibid., p. 191.

[87] The Revd. Alfred Barrett, *Holy Living; Exemplified in the Life of Mrs Mary Cryer*, London, 1845, pp. 4, 9.

and spiritual vacillation. 'I have no business upon earth', she proclaimed, 'but to be a vessel of His grace, an instrument for His glory'. And later: 'If ever Satan desired to set me as a mark for his fiery darts, it is now; he generally uses some human agent for his bow, and his arrows come hot from the infernal quiver.'[88] Marriage to a Methodist missionary, Thomas Cryer, relieved much of her spiritual 'groaning'. But it was in her love of fellow sufferers that she found the most 'sin-consuming, self-annihilating, Christ-exalting, perfect, burning love'.

> I find the cross in canvassing from door to door for Missions, and perhaps not succeeding in one case in twenty; I find it still more in begging for a poor starving fellow-creature, and perhaps now and then meeting with a chilling repulse; . . . but, most of all, I find it in going from door to door on the visiting plan, trying to persuade sinners to attend God's house, and flee from the wrath to come. . . . But O, when I have to make my own way, and meet the cold looks and even the rude rebuff of those who will not be subdued by kindness and courtesy, then nature does shrink; . . . Yet I dare not give it up: it is God's work; He is with us in it, and we have some little fruit. But O, I want more love for Jesus, more courage in His cause, more zeal for God, more melting pity for sinners, more vehement desire to snatch these firebrands from the flame. I often think I ought not to be reckoned among the soldiers, till I burn with a nobler, warmer, bolder spirit, for the fulfilment of His love's redeeming plan.[89]

Mary Cryer died at Manargoody Mission, India, aged thirty-three.

From the writings of women like Jane Gibson and Anna Clarke, Margaret Burton and Mary Cryer, and a great many others, we can see that benevolence was not simply, as evangelical doctrine would have it, the natural result of conversion, a product of a true acceptance of the Gospel covenant. It was often a product of that anxiety of soul which asks, am I saved? The Scriptural passage 'I was sick, and ye visited me' was, among others, a powerful influence on these women; and the text that followed, if taken literally, could be unsettling: 'Depart from me, ye cursed, into the eternal fire which is prepared for the devil and his angels: for I was an hungred, and ye gave me no meat: I was thirsty, and ye

[88] *The Devotional Remains of Mrs. Cryer*, London, 1854, pp. 43, 63–4.
[89] Ibid., pp. 102–4.

gave me no drink: I was a stranger, and ye took me not in; naked, and ye clothed me not; sick, and in prison, and ye visited me not.' (Matthew 25: 41–3.) Women whose hearts 'yearned for sinners' visited the poor not only in anticipation of grace but in fear of damnation. The heart-burnings, the impatience, the weeping and supplication so common among visitors were not signs of salvation. These were women in need, a need so great that in their visits they sometimes confused shadow for substance, fabricated conversions, and meddled with the spiritual lives of their subjects. The doubts which made them long to be converted in the first place lingered on after they believed their conversions to have taken place. And so when they harangued backsliders they were also looking into their own souls. The opening lines from one woman visitor's diary sums up this uneasiness of mind: 'Do I wish for Heaven really? Do I fear Hell?'[90]

The zeal to save souls, so much a part of evangelical benevolence, was never more conspicuous than during visits to the dying. Take Mary Timms, from Somerset, who specialized in deathbed conversions. When word reached her that a neighbour was failing she gathered her improving books and set off post-haste. As her biographer relates, 'her solicitude increased in proportion as she saw persons approaching another world'. Once she discovered the sick person's views of immortality, she 'instructed, exhorted, or warned, as the case required'.[91] As her own spiritual journey was full of stops and starts—her diary is filled with religious doubts and a recurring death-wish—she was hardly a good spiritual guide for those who wished to die in peace. But then her needs were as great in their way as those of the dying she must, on occasion, have

[90] From an anonymous mid-nineteenth-century manuscript diary, in the author's possession. During her arduous intellectual journey from Christianity through atheism to theosophy Annie Besant also 'found some relief from the mental strain in practical parish work, nursing the sick, trying to brighten the lot of the poor'. *An Autobiography*, London, 1893, p. 104.

[91] *Memoirs of the late Mrs. Mary Timms*, ed. the Revd. E. Morgan, London, 1835, pp. 27–8. See also *The Extraordinary Conversion and Religious Experience, of Dorothy Ripley*, London, 1817, pp. 1–3. The practices of such women should be compared with the advice given to visitors of the aged, for example in the *Churchman's Family Magazine*, ii, 1863, pp. 429–35.

tormented. Like so many evangelical visitors, Mrs Timms was anxious to reach the Garden of Gethsemane, so anxious that she would dash herself against the gates. She died of smallpox, aged twenty-six, soon after her marriage to a Methodist minister. Her seven-month-old child died a few weeks before her.[92]

Death is everywhere present in such memoirs, and it gave great stimulus to religious and charitable practice. Death was, of course, experienced more directly then than today, in all its horror. Few families could hope to be spared from a serious illness, or a sudden death, especially of a child (about one in seven children died as infants in the late nineteenth century).[93] The resulting emotional distress was often at the root of evangelical conversions. As it has been suggested, the fear of a child's dying before 'finding Jesus' was one of the reasons why evangelical parents sought so desperately to achieve early conversions in their children. Margaret Burton, who, as we have seen, lost her father when she was four, lost her mother when she was nineteen and later one of her daughters. Her daughter Mary saw her sister's death and later those of her mother and father, all before she was twenty-five. Similar cases abound.

Women saw much more of disease and death than men, and it concentrated their minds, not least on good works. The practice of Christian benevolence made it somewhat easier to reconcile themselves to death, including their own; benevolence, it should be remembered, was commonly taken as a sign that they may be spared from eternal torment. Women commonlv took on a new piety after witnessing a death. As a girl Ellen Ranyard was so moved by the loss of a friend that it reinvigorated her interest in visiting the poor.[94] The charitable Georgina King Lewis, the ninth child of a mid-century evangelical Quaker family, saw several of her brothers and sisters die as a girl, events which 'called forth [her] sympathy in later years towards those who were

[92] *Memoirs of the late Mrs. Mary Timms*, pp. 87–93.

[93] B. R. Mitchell, *Abstract of British Historical Statistics*, Cambridge, 1971, pp. 36–7.

[94] *Dictionary of National Biography*.

bereaved'.[95] Josephine Butler's experience at the time of her daughter's death followed a similar pattern. As often as not, the sympathy aroused by disease and deathbed scenes had an impersonality about it. The suffering that women had to witness, and most endured in childbirth, seems to have produced a certain toughness in them. They may have become inured to it by experience.

With this departure into the subject of death and dying, we have lost sight of some of the visitor's most urgent motives. For many volunteers, especially those without families, the daily or weekly round among the poor was a source of immense pleasure. Despite encounters with hardship and disease, it could be the one bright spot in an otherwise dreary landscape. 'You would scarcely believe how much I look forward to my half hours with the poor', remarked May Percival, one of the visitors in Isabel Reaney's novel *Waking and Working*. 'It is not because we are better than they are that we seek to do them good', she added, quoting the words of her friend Grace Sullivan, 'but because we are so conscious of our own need that we are able to sympathize with theirs'.[96] To be needed, to be counted upon, to be called 'dear' and 'friend' by those more obviously in distress was a great reward. 'The children know me, and speak my name', wrote Anne J. Clough in her journal after an outing with the poor (she went on to become the first Principle of Newnham College). 'This was delicious to me, and worth more than a thousand praises.'[97]

Such experiences made up for the rebukes, the heartaches, and the doors slammed in the face. To be known and looked up to by one's neighbours, however humble, gave women visitors stature in the community. Even if only voluntary work, it gave them a status in short supply elsewhere. Moreover, it broke the domestic routine, got them into the neighbourhood, and not least of all it gave them an intimate

[95] *Georgina King Lewis, an Autobiographical Sketch*, ed. Barbara Duncan Harris, London, 1925, p. 25.

[96] Mrs G. S. Reaney, *Waking and Working*, London, 1874, p. 158.

[97] Blanche Athena Clough, *A Memoir of Anne Jemima Clough*, London, 1897, p. 22, quoted in Margaret B. Simey, *Charitable Effort in Liverpool in the Nineteenth Century*, Liverpool, 1951.

knowledge of local affairs. Such knowledge, for instance, could be very useful in bringing together girls desperate for work in domestic service and potential employers. It could also be an effective check on truancy among schoolchildren. The visiting of women was not then simply a Christian duty, though that was a good part of it; it represented basic human urges: to be useful, to be recognized, to be informed, to be diverted, to 'keep up with the Joneses', to gossip, to be loved.

Guilt, at least class guilt, does not appear to have been a very powerful motive in visiting, or in the charitable work of women generally. Most nineteenth-century philanthropists held a hierarchical view of society and assumed that distinctions between rich and poor were God-given and likely to persist. Few of them asked how a God of love could have created a world in which social divisions could be so often cruel and unjust. Being practical they sought to alleviate the worst abuses of society without undermining their authority. The opportunity to be charitable, after all, depended on social inequality. This is not to say that they felt no guilt at all; but for women it was more likely to be inspired by Eve and not class advantage. 'We brought all the sin into the world', said the writer and temperance worker Clara Lucas Balfour, 'involving man in the ruin that he was not the first to seek, and it is the least that we can do to offer him a little good now'.[98] The world can be very agreeable for the privileged if they leave it alone, but religious sensibilities made this impossible for many nineteenth-century women.

To impose some discipline on visitors with such varied and often complex motives was a constant problem for the visiting charities. It was no easy task to get women who were anxious to save sinners to master the public health statutes and the Poor Law. They commonly thought that temporal relief interfered with their spiritual message. Other visitors thought that spiritual guidance alone was little solace to a family with nothing on their table. And others still thought that to carry a Bible in one hand and a food ticket in the other was not an altogether happy compromise. But then how was relief best administered? More to the point, how

[98] Clara Lucas Balfour, *Working Women of the last Half Century: the lesson of their lives*, London, 1854, p. 169.

was it best administered in the poorest districts, where funds and volunteers were in short supply? How could the health and happiness of the poor best be promoted? And what were to be the respective spheres of charity and the Poor Law in this process? These were among the issues at the heart of visiting, which the societies, by one means or another, tried to resolve. The guide-lines and handbooks which they recommended were, as we have seen, devices used to improve the performance of visitors; but there was a constant need for experimentation, especially in the administration of relief. Among the most interesting experiments, though not now the best known, were made by women.

Some of the most imaginative ideas on the difficulty of penetrating the poorest neighbourhoods came from the Low-Church Anglican Ellen Ranyard.[99] Born in Nine Elms in 1810, the daughter of a cement-maker, she had from her earliest years been a visitor of the poor. The female poor aroused her special concern, and it was during a summer walk through St. Giles in 1857 that 'the *misery of our sisters* there . . . brought forth the idea of the Bible-woman'.[100] This Bible-woman was to be a 'native agent' at home in the courts and alleys, who would not excite the pride or the hope of the residents. Her job was to be a seller of Bibles and an adviser on domestic matters. She was, in Mrs Ranyard's words, to be the 'missing link' between the poorest class and their fellow men, the agency by which society's outcasts were to be brought back to morality and religion. The first Bible-woman, recommended by a City Missionary, was Marion Bowers, a devout woman in her thirties who had been orphaned as a girl and who earned her small living cutting fire-papers and moulding wax flowers. For six months she was given a trial, her small salary of ten shillings a week paid for by a grant from the British and Foreign Bible Society. Armed with a bag of Bibles she 'plunged' into the 'most

[99] There is no scholarly treatment of Ellen Ranyard. For a brief sketch see the *Dictionary of National Biography*; also Elspeth Platt, *The Story of the Ranyard Mission, 1857–1937*, London, 1937; Rose Emily Selfe, *Light amid London Shadows*, London, 1906. There is considerable manuscript material relating to Mrs Ranyard's work in the Greater London Record Office.

[100] [Ellen Ranyard], *London, and Ten Years Work in It*, London, 1868, p. 7.

appalling recesses' of Soho. And despite the rebukes, threats, and 'a bucket of filth' thrown on her, she managed to sell, on instalment, seventy Bibles by the end of the first month.[101]

Mrs Ranyard had inaugurated not only a 'women's mission to women' but also the first corps of paid social workers in England.[102] In 1867, ten years after the initial experiment, there were 234 Bible-women in London's poorer districts, and Mrs Ranyard's Bible and Domestic Female Mission, as it was then called, had raised over £133,000. Most of this came from contributions, but some of it, over £8,000, came from grants from the British and Foreign Bible Society, which saw the Bible-women as its London agents.[103] Other small grants came from the Society for the Relief of Distress, which had a policy of assisting projects in poor neighbourhoods.[104] It was always Mrs Ranyard's intention to work with other societies, to play down sectarian rivalries, and it paid off. By 1867, she was able to pay salaries of about £32 a year to her Bible-women, a figure which, though modest, compared favourably with moulding wax flowers or the lower forms of domestic service. There were plenty of applicants; the successful ones received a three-month course in Scripture, the Poor Law, hygiene, and in later years the principles of the Charity Organisation Society. They were then assigned to one of the many districts, preferably the one in which they lived, that stretched from Highgate to South Lambeth, West Ham to Brentford. The scheme spread rapidly to other English cities; by 1862 there was 'some description of Bible-woman . . . in almost every town in England'.[105] In time they were

[101] Ibid.; *The Quarterly Review*, cviii, 1860, pp. 7-16.

[102] Kathleen Heasman, *Evangelicals in Action*, London, 1962, p. 37. From its inception the London City Mission used paid visitors, but they were prohibited from giving temporal relief. Mrs Ranyard was a faithful subscriber to the London City Mission and its work probably influenced her own visiting schemes.

[103] [Ranyard], *London, and Ten Years Work in It*, pp. 8-10, 13.

[104] See the *Minute Book of the Executive Committee of the Society for the Relief of Distress*, A/SRD/1/2, *passim*, the Greater London Record Office.

[105] Ellen Ranyard, *The True Institution of Sisterhood: or a Message and its Messengers*, London [1862], p. 16. In London, the Parochial Mission Women's Association, founded in 1860, worked along the same lines in some of the same neighbourhoods, under the supervision of the parish clergyman. See *Meliora*, xii, 1869, pp. 89-91.

all over the world, from North America to India, Burma to Australia.

From the earliest years, Mrs Ranyard instructed the Bible-women to advise the poor on household management; like so many others she believed that the training of wives and daughters in cooking, cleaning, and needlework was essential to improving the condition of the poor. The Bible-women quickly discovered that when they gave such assistance it was easier to sell Bibles. And before long they collected subscriptions to pay for clothing, coals, food, and furniture as well. 'Nothing is valued but what is paid for', argued Mrs Ranyard, and so she established the rule that goods had to be paid for on delivery, except in emergencies.[106] The success of the various provident schemes was phenomenal. In the first ten years Bible-women collected over £44,000 from the poor, much of it for clothing and furniture.[107] Mrs Ranyard could say that the poor spent this great sum not only on the material necessities of life, but also on providence and self-control, self-reliance and self-respect. People with property, she might have added, were respecters of property.

The administration of the Bible and Domestic Female Mission was an ingenious mixture of paid and volunteer, working-class and middle-class workers respectively. Mrs Ranyard found it necessary and desirable to enlist 'lady superintendents', who met weekly with individual Bible-women to read their reports, exchange information on cases, pay salaries, and run the mothers' meetings held in the mission rooms set up in each district as retreats for the poor. A division of duties emerged. The Bible-women concentrated on Bible sales and giving tips on domestic economy; their superintendents concentrated on administration and, more conscious of the dangers of indiscriminate relief, on the material welfare of the cases referred to them. The Mission instructed both classes of women to show no sectarian bias in their work. Members of all faiths, and perhaps more to the point, none, were to be encouraged to attend the district mission, and, whenever possible, put in touch with local

[106] Quoted, *The Quarterly Review*, cviii, 1860, p. 15.
[107] [Ranyard], *London, and Ten Years Work in It*, pp. 8–10.

churches. This was no small task in the many London districts short of clergy, in which church attendance on Census-Sunday 1851 was commonly less than 15 per cent.

The problems associated with sick visiting were every bit as profound as those of religious instruction. Mrs Ranyard's writings are full of the most distressing case-studies of the sick and dying. The Bible-women, for all their ingenuity, were unfit to give anything but the most rudimentary medical advice. In touch with nurses and nursing schemes, including those of Pastor Fliedner at Kaiserswerth, Mrs Ranyard decided to try another experiment. She may have been influenced by Agnes Jones, who in 1861 worked with the Bible-women in London, and who went on to do important work in the mid-1860s in the training of pauper nurses at the Liverpool Infirmary. When a gentleman donated a building suitable as a centre for nursing in 1868, Mrs Ranyard quickly put her plan into action. She signed up a staff of poor women to train as itinerant nurses, to complement her corps of Bible-women in the slums. In the first seven years seventy-eight women were trained, mostly in the general and lying-in wards of Guy's Hospital. In 1868–9, the first recruits, who were paid £39 a year, made 5,000 visits to ninety-nine patients. The figures rose to 111,601 visits to 4,392 patients in 1874.[108] Along with the work of the smaller East London Nursing Association, also established by women in 1868, Mrs Ranyard's programme was the earliest example of district nursing in London. It followed William Rathbone's Liverpool nursing scheme by nine years. Her Bible-nurses, as she called them, were familiar figures in London's poorest neighbourhoods. With Florence Nightingale's *Notes on Nursing* in one pocket and the Scriptures in the other, they gave medicines and solace to patients with ailments ranging from bedsores to 'sloughs of despond'. Their work has been called 'a gallant pioneering achievement'.[109]

One of those impressed by Ellen Ranyard's ideas was Octavia Hill, who was to become well known for her work in housing and the Open Space movement. 'It is wonderful

[108] [Ranyard], *Nurses for the Needy*, pp. 118, 250.
[109] Mary Stocks, *A Hundred Years of District Nursing*, London, 1960, p. 25.

what [the Bible-women] have done', she wrote to a friend in 1859, 'they give nothing away, but get people to buy beds and clothes, for which they pay gradually'.[110] Appalled by the waste and unhappy effects of charity badly administered, schemes of providence appealed to her. Like Mrs Ranyard she wished to help the poor make their own way in the world without the deadening effects of indiscriminate almsgiving. And like Mrs Ranyard her approach was simple and uncompromising. She saw the source of social distress in individual character, not in the structure of society. The remedy was to be found among the poor themselves, in their acceptance of individual responsibility. Though historians have tended to play it down, the inspiration for her views came directly from Christianity.[111] 'I do not feel that in urging any of you to consider the right settlement of questions of temporal relief', Miss Hill said to a group of district visitors, 'I am asking you to devote yourselves to a task which is otherwise than holy'.[112] 'The foundation of all charity', she asserted, was the life of Christ.[113]

Octavia Hill, encouraged by the financial support of John Ruskin, began to put her ideas into practice in the 1860s. In 1870 the Reverend W. H. Fremantle, Rector of St. Mary's, Bryanston Square, who had established the first district committee of the COS in Marylebone in 1869, persuaded her to take over the relief work in the poorest neighbourhood under his care, the Walmer Street District. Her plan was straightforward: she wished to perfect the administration of relief in order to put people to work. The substitution of employment for doles had several advantages in her mind: it did not impoverish anyone; it was educational; and it encouraged self-respect. Miss Hill was a founder of the COS and a member of its Marylebone District Committee,

[110] *Life of Octavia Hill as told in her letters*, ed. C. Edmund Maurice, London, 1913, p. 171.

[111] David Owen, for example, emphasizes Thomas Chalmers' influence on her charitable views, but Chalmers only reinforced her already powerful Christian attitudes.

[112] Octavia Hill, *District Visiting*, London, 1877, p. 14, reprinted from *Good Words*.

[113] Octavia Hill, *Our Common Land (and other short Essays)*, London, 1877, p. 61.

which referred applicants for relief to local charities; and now she joined the Marylebone Relief Committee based in St. Mary's church. To facilitate her work she acted as liaison officer between these two committees and between the committees and her staff of visitors, initially thirty-five women. This figure doubled in 1874, at which time she was even able to recruit a few men.

She had nothing against using volunteers. Indeed, she believed that women with family responsibilities made the best visitors: 'Depend upon it, if we thought of the poor primarily as husbands, wives, sons, and daughters, members of households, as we are ourselves, instead of contemplating them as a different class, we should recognise better how the hours training and high ideal of home duty was our best preparation for work among them.'[114] Most of her volunteers, despite being wives and mothers, needed training. Miss Hill thus set up a centre for visitors, where instruction was given and information exchanged. She encouraged her visitors to work for local charities, with school-boards and the Charity Organisation Society. The more hats they wore the better, for they would then be able to see the poor from more than one perspective. All of them kept a visitor's book, which served the dual purpose of keeping them up to the mark and provided accurate information on cases. There was nothing new in this, but Miss Hill carried it one step further. In 1873, her visitors began to exchange information on cases with Poor-Law officials in Marylebone.[115]

Thus household visitors, perhaps for the first time, established a close working partnership with Poor-Law officials. Each day Miss Hill received a list of the applicants for relief from the officer in charge of St. Mary's Poor-Law District. She handed this information on to the visitors, along with a

[114] Octavia Hill, *District Visiting*, p. 6.

[115] Octavia Hill, *Letter accompanying the Account of Donations received for Work amongst the Poor during 1872*, London, 1873, p. 2; Octavia Hill, *Homes of the London Poor*, London, 1875, pp. 147–56; Fremantle, *St. Mary's, Bryanston Square*, pp. 29–39. Miss Hill's activities in Marylebone should be seen in the light of the system of visiting used in Elberfeld, Germany; see *The Poor Law System of Elberfeld* [London, 1870]; *P. P., Reports from Commissioners, Third Annual Report of the Local Government Board*, 1874, xxv, pp. 126–30.

blank form, on which they were to gather facts useful to the Guardians. It was knowledge that Guardians would have difficulty getting elsewhere; for one thing it included the amount of aid the applicant was receiving from local charities or from the visitor. This process was to give a clearer definition to the relationship between private and public relief, the gift and the rate, which so often had been blurred in the past. As Miss Hill and her co-workers in the COS would have it, charity was to assist deserving cases, those who could be helped by preventive and remedial action; the Poor Law was for the undeserving destitute. There should be little, if any, overlapping of authority; charity was to begin where public relief ended. This plan was to result in a decline in pauperism, increased frugality in the working classes, and a drop in the poor-rate. It was not a system that lived up to expectations, for reasons which have been discussed at length elsewhere.[116] But it did give visitors, who were at the very heart of the scheme, a new dimension and thereby a better appreciation of the problems of poor relief. Under Miss Hill's guidance they were turning into something resembling the modern social worker.[117]

These brief sketches of the district-visiting schemes of Ellen Ranyard and Octavia Hill were meant largely to illustrate that visiting was becoming increasingly sophisticated as the nineteenth century progressed and that women played an important role in this process. The history of visiting is in part a history of its transition from amateurishness to sophistication, the grafting of social science methods on to religious precept and church organization. But we must not overlook the fact that the expansion and refinement of visiting methods were directly related to the growth of government, particularly at the local level. Two examples will illustrate this: visitors willingly referred cases to medical officers of the local Boards of Health created by the Public Health Act

[116] See Charles Loch Mowat, *The Charity Organisation Society, 1869–1913*, London, 1961, Chapter VI; Owen, *English Philanthropy, 1660–1960*, Chapter VIII; Jones, *Outcast London*, Chapter XIII.

[117] Hill, *Homes of the London Poor*, pp. 150–1; Octavia Hill was later active in Southwark with the women's University Settlement Committee. She provided a description of that society's sophisticated visiting practices in *The Nineteenth Century*, xxx, 1891, pp. 164–9.

of 1848. Indeed, the distinguished Ladies' Sanitary Association, one of the great sanitary societies, distributed information on the laws of health to cottagers and artisans.[118] And when the Education Act of 1870 established school-boards, the visiting societies put their volunteers on the alert to discover and report the whereabouts of school-age children. In short, as the agencies of government expanded, the duties of the visitor tended to enlarge as well. But non-governmental agencies were part of the process too, and not only the philanthropic societies. Many of those who listened to lectures on public health, the Poor Law, and visiting itself at the social science congresses and such institutions as the Working Men's College, Red Lion Square, were visitors. By the last quarter of the century many of them were becoming less sectarian and more professional, in other words more and more alike, even though most of them were still volunteers.

Women like Octavia Hill and Ellen Ranyard played a part in this transition 'from charity to social work', but we must not overstate their contribution to 'social science', at least as it is understood today. Being untheoretical, they sought a more systematic organization of charity rather than a fuller comprehension of the social process. Few besides Beatrice Webb moved beyond social reform to an analysis of social structure. Women did turn out in large numbers at social science congresses (one critic thought they dwelled overmuch on the 'social'),[119] but to mid-Victorians social science was largely practical and piecemeal. It sought to identify and sub-divide social problems and gather information to remedy them.[120] Not surprisingly then, even advanced charitable women had only a vague notion of society's economic organization and had little knowledge of social theory. Their memoirs suggest that such matters did not much interest them, for they rarely came within the scope of their education and experience.

[118] See the Reports of the Association; *The English Woman's Journal*, iii, 1859, pp. 73–85; and W. C. Dowling, 'The Ladies' Sanitary Association and the Origins of the Health Visiting Service', M. A. Thesis, University of London, 1963.

[119] *Blackwood's Magazine*, xc, 1861, pp. 463–78.

[120] See E. M. Yeo, 'Social Science and Social Change: A Social History of some Aspects of Social Science and Social Investigation in Britain 1830–1890', Sussex Ph. D. Thesis, 1972.

Their philanthropy, which often consisted of short-term practical solutions to immediate problems, reflected the pragmatic, unanalytic mentality encouraged in the other spheres of their lives.

While having little taste for theory, many women visitors nevertheless had a strong sense of their aims, which commonly included the abolition of pauperism, the promotion of Christian family life, and the encouragement of self-help among the poor. Those of them who thought about it at all were also likely to want to restrict the government's role in poor relief, for they could hardly avoid the conclusion that female influence was likely to decline in a system of relief run by the state. Perhaps a few of them also concluded that social science theories were inclined to promote state intervention. Be that as it may, the aims of the more enlightened charitable women were not very different from those of the leading philanthropic thinkers of the nineteenth century, for example Thomas Bernard and the Scot Thomas Chalmers. Certainly some women came under the influence of such authorities, Octavia Hill for one. But many other women of a benevolent disposition were working along parallel lines without ever having heard of Bernard or Chalmers. If they read anything at all on social and philanthropic schemes, it was more likely to come from the pens of the numerous women writers, like Mary Bayly, Mary Carpenter, or Octavia Hill herself, who shared their special concerns and spoke their language. Not a few of them would have been surprised, probably alarmed, to discover that philanthropy was increasingly on the defensive against the attacks of political economy and social theory. Their political isolation and the dictates of need and religion removed most women from such issues.

District visitors and the societies they represented were rather more sensitive to the debates of political economy and social science than people engaged in other forms of philanthropy, foreign missions for example. Enlightened visitors were well aware that charity was often unjust; it raised money, not necessarily from those who could afford it, but from the compassionate and weak; it distributed it in turn, not necessarily in proportion to need, but often to the dishonest and undeserving.[121] As the critics warned, the casual almsgiving of visitors could be an unproductive use of capital, which fostered indigence, improvidence, and hypocrisy in the poor. The story of a Yorkshire beggar illustrates their point. When asked by a lady whether he could read or write he replied, with a grin: 'No Ma'am, I can't, . . . and if I'd known as much when I was a child as I do now, I'd never have learnt to walk and talk.'[122] Visiting did produce mischief, and damage of a kind that could become permanent. Did the good conferred by visitors, which was often of a temporary nature, make up for it? The answer in the nineteenth century, Harriet Martineau and her ilk notwithstanding, was a resounding yes. The scale of the contributions and of the recruitment to the visiting societies was proof.

For want of a better system of ameliorating human misery, the visiting charities pushed on. And despite the abuses, the cant, and the danger of making paupers of those they relieved, they had ample evidence of the benefits of their work. They could argue that institutions like the COS and the Ranyard Mission were making visiting more efficient and professional; they could argue that self-help was the guiding principle of an increasing number of visiting societies; they could argue that the poor themselves were sometimes in the forefront of the movement as visitors; and they could add that when they distributed alms it was usually to people who were already paupers, for reasons other than charity. Moreover, they had every reason to be proud of the countless mercies shown to the victims of accidents and infectious diseases, to the lonely, to the aged, to pregnant women, to starving children. Charity could not pauperize a starving child and the dead are rarely hypocrites.

As we have seen, it is sometimes difficult to determine who received the greater benefit, the visited or the visitor. 'The poor teach and help us quite as much as we can teach and help them', aptly remarked one visitor from fiction.[123] Or as Elizabeth Barrett Browning wrote:

[121] *The Westminster Review*, cxxxv, 1891, pp. 382-3.
[122] Ibid., p. 373.
[123] Reaney, *Waking and Working*, p. 158.

Thy love
Shall chant itself its own beatitudes
After its own life-working. A child's kiss
Set on thy sighing lips shall make thee glad;
A poor man served by thee shall make thee rich;
A sick man helped by thee shall make thee strong;
Thou shalt be served thyself by every sense
Of service which thou renderest.[124]

Cynics might argue that the poor are with us so that the rich may be virtuous. The uncynical Wesley put it another way. He believed that Adam's fall offered man a chance to save his soul, that suffering humanity gave man the opportunity to exercise his capacity for good works.[125] In a fallen, iniquitous world, the poor could teach the visitors the art of sympathy and the path to heaven; the visitors could teach the poor the facts of domestic economy and the path to heaven.

One effect of the pervasive belief in original sin, or the 'natural depravity' of man as Wilberforce put it, was that it made charitable labour somewhat more egalitarian than it might have been otherwise. Differences of rank had less meaning to those who believed that in the sight of God all men were equally sinners. Some visitors, of course, were much better at playing down the distinctions between themselves and those whose homes they entered; yet all were encouraged to do so. 'Take the love of Christ for your motive, *and be natural*', wrote one churchman, 'deal with the poor as you would yourself be dealt with by another'.[126] 'In visiting the poor it is very desirable to aim at forgetting as far as possible, difference of rank', remarked one female visitor.[127] Another reminded her readers that 'the lower ranks of life are equally the care of the Eternal Parent'.[128]

[124] 'A Drama of Exile', *The Poetical Works of Elizabeth Barrett Browning*, 6 vols., London, 1903, i, pp. 84–5.

[125] *The Works of the Rev. John Wesley*, vi, pp. 236–7.

[126] Briggs, *Chequer Alley*, p. 11.

[127] [Charlotte Bickersteth Wheeler], *Memorials of a Beloved Mother: Being a Sketch of the Life of Mrs. Cooper*, London, 1853, p. 22.

[128] *Extracts from the Journal of the late Margaret Woods, from the Year 1771 to 1821*, Philadelphia, 1850, p. 141.

Bishop Thorold's advice to the lady supervisors of the Bible-women went further: 'you should never let [the Bible-women] feel there is a gulf between your position and their own. It may happen that thirty years hence they may occupy a high place in the Kingdom of God, while you are in a lower one.'[129] Egalitarian beneficence, kindled and extended by the visiting of the poor themselves, worked to fulfil one of visiting's great objects: 'reuniting the severed sympathies' of the nation's classes.[130]

By their labours the women visitors made a considerable contribution to the stability of nineteenth-century English life; at the same time their own lives were not unaffected. Visiting was an activity that a vast number of women had in common, from great ladies on their estates to working women in the slums. It gave women a shared experience, a wider sympathy, and first-hand knowledge of poverty and distress which were rare among men except those of the labouring classes. As volunteers and as women, they were badly placed to do much more than administer temporary relief. But as more and more of them filed out of their homes to take up their charitable rounds, they pushed for the opportunity to put their compassion and experience to wider and wider use. By the mid-century, women visitors had helped to change the temper of English society, while strengthening their own lives and raising their expectations. As Lucy Aikin wrote to Dr Channing in 1841: 'The practice of superintending the poor has become so general, that I know no one circumstance by which the manners, studies and occupations of Englishwomen have been so extensively modified, or so strikingly contradistinguished from those of a former generation.'[131] For the women visitors the maxim 'to help the poor to help themselves' had an ambiguous meaning.

[129] Platt, *The Story of the Ranyard Mission*, p. 49.

[130] [William Rathbone], *Social Duties*, London, 1867, p. 14.

[131] *Correspondence of William Ellery Channing, D. D. and Lucy Aikin*, p. 397, quoted in Ian Bradley, *The Call to Seriousness*, London, 1976, p. 125.

V

In Public and Charitable Institutions

The door-to-door visiting carried out by the women of England was an important innovation, but despite its merits there were many people who remained little affected by it, especially those without homes to be visited. Among others, the inmates of prisons, workhouses, orphanages, hospitals, refuges, and asylums were not susceptible to the influence of household visitors. Thus women pioneered and promoted a variety of institutional schemes which complemented their neighbourhood rounds. Whether they managed institutions or 'visited' them their aim was the same: to bring the nation's sick, poor, and outcast into contact with their more fortunate neighbours, who would call forth their better feelings and lead them back to the fold. Often against considerable opposition women set up their own societies or persuaded male managers of public and charitable institutions to give them the opportunity to serve. The expansion of their activities, as we shall see, was gradual, and not without set-backs to the cause. But in most institutions women came to wield power and produce benefits.

There was as much variety in institutional visiting as in household visiting. Casual visitors, usually women, received permission to work in particular hospitals, infirmaries, workhouses, asylums, and prisons. And large, highly organized societies sought to get their agents into every workhouse, or prison, or medical charity in the country. Many of the societies were small local associations of women, which sprang up to look into the affairs of an institution in the neighbourhood. Some of them practised district visiting as well as institutional visiting, and most were in touch with other charities which complemented their work. As in household visiting there was sectarian rivalry, which was particularly heated between evangelicals, who were avid institutional visitors, and Catholics.[1] Virtually every sect or religion was represented. There were Catholic ladies' societies for visiting hospitals, infirmaries, and workhouses, which normally restricted their labours to giving assistance to Catholic inmates.[2] There was a Jewish prison-visiting charity which assisted only Jews.[3] The Quakers were also active; Mrs Fry's work being, of course, the most widely known. But whether High Church or Low, Dissenter, Catholic, or Jew, the visitors and institutional visiting societies shared at least one fundamental belief: that contact between inmates and the benevolent was wholesome and essential.

The origins of women's institutional work are obscure. We know something about early female-managed societies which dealt with poor and outcast inmates, but it would be impossible to put a date on the first female visit to a hospital, workhouse, or prison.[4] More precision is possible with individual institutions, for memoirs and the reports of charitable societies sometimes tell us when women volunteers first entered a particular hospital, workhouse, or prison. The advance of women into such institutions went hand in hand with their advances into the nation's homes. Institutional and household visiting, and for that matter brothel, barracks, or lodging-house visiting, had much in common, and many volunteers worked in more than one sphere. Whichever sphere a woman chose to make her speciality, she was responding to a call on female benevolence which was sweeping in its implications. 'Women are prohibited from public service of their country by reason and decorum', wrote

[1] For an example of the problems this rivalry posed for workhouse and prison authorities see Public Record Office, HO 45 OS6840.

[2] *Handbook of Catholic Charitable and Social Works*, p. 15. *The Catholic Directory, Ecclesiastical Register, and Almanac, for . . . 1880*, London, 1880, pp. 319–20.

[3] *Fifth Annual Report of the Jewish Association for the Diffusion of Religious Knowledge*, London, 1865.

[4] There is an early example of prison visiting quoted from Wesley's *Journals* in R. S. E. Hinde, *The British Penal System, 1773–1950*, London, 1951, p. 19. A woman named Sarah Peters entered Newgate in October, 1748, 'sometimes alone, sometimes with one or two others, visited all that were condemned in their cells, exhorted them, prayed with them, and had the comfort of finding them, every time, more athirst for God than before'. She died of gaol fever (typhus) in November, 1748.

Priscilla Wakefield in 1798, 'but they are not excluded from promoting its welfare by other means, better adapted to their powers and attainments'.[5] She, like a growing number of her contemporaries, interpreted these 'powers and attainments' rather broadly. Certainly the visiting of workhouses, schools, refuges, prisons, and other institutions for the improvement of morals were fit and proper concerns for women volunteers.

This was not a view shared by everyone. There was relatively little opposition to women household visitors, largely because no one asked the poor for their permission to be visited. Public and charitable institutions were different, for their governors and managers had to be convinced that female volunteers were a good thing. Many of them agreed with the commentator in the *Anti-Jacobin Review*, who complained of women wandering about the kingdom prying into prisons when they should be at home disciplining their children.[6] Others, with an instinctive dislike of innovation, recoiled in horror at the idea of allowing ladies to witness the scenes that commonly took place behind bars and in the sick wards. Still others believed that women were incapable of working with their staff with the required prudence and discretion; they feared meddling interference. And all too many feared that female busybodies might expose abuses in their institutions. This could prove not only embarrassing to the authorities personally but might lead to a call for reforms for which there was little money. The religious issue also emerged, for managers, governors, and chaplains were chary, with justice, of letting earnest sectarians loose on their inmates. Those who sought reasons to exclude women visitors found them.

How was such prejudice or self-interest to be overcome? Women initially maintained, though some were being disingenuous, that they wished only to be given access to the female wards of the various institutions, where they would introduce habits of order, a sense of propriety, and help to oversee the work of matrons and nurses. They defended themselves by saying that there were many things relating

[5] Priscilla Wakefield, *Reflections on the present condition of the female sex; with suggestions for its improvement*, London, 1798, pp. 82–4, 95–6.

[6] *The Anti-Jacobin Review*, lviii, 1820, p. 552.

to domestic management that men found strange and disagreeable. The familiar leap was made from the home to the institution: 'A lady visitor in an hospital or Asylum, should be to that institution what the kind judicious Mistress of a family is to her household,—the careful inspector of the œconomy, the integrity and the good moral conduct of the housekeeper and other inferior servants.'[7] The mother, the sister, and the daughter, it was invariably argued, had a special blend of sympathy and experience that the country's institutions could not do without. At the side of the ill and dying, remarked Frances Power Cobbe, was 'the very place which the sternest contemners of the sex declare to be a woman's proper post'.[8] Were they to be denied the opportunity to take it up?

No, but they did not always get their way by reasoned argument. As often as not it was the persuasiveness of their money or their special influence as well-placed wives or daughters which carried the day. One of the first hospitals to adopt a ladies' visiting-plan is a case in point. The Unitarian philanthropist, Catharine Cappe, one of the pioneers of the movement to get women visitors into public and charitable institutions, relates how she had campaigned for years, unsuccessfully, to induce the York County Hospital to open its doors to women visitors. When reason failed she resorted to the purse. In early 1813, the hospital was suffering from a shortage of funds. She contacted a large number of wealthy women in Yorkshire and persuaded them to become annual subscribers at two guineas each, on the express condition that female visitors would be appointed. By the end of the summer she had received favourable replies from over fifty of the county's most distinguished women, including the wives of the Archbishop, the High-Sheriff, and the Members of Parliament. She forwarded her proposal to the Hospital Governors and they came around to her terms. By the end of 1813 two women were visiting the sick

[7] Catharine Cappe, *Thoughts on the Desirableness and Utility of Ladies visiting the Female Wards of Hospitals and Lunatic Asylums*, London, 1816, p. 376.

[8] *Macmillan's Magazine*, iii, 1861, p. 456. See also *The Nineteenth Century*, xxviii, 1890, p. 952.

wards on a regular basis.[9] The following year the York Lunatic Asylum, which was founded in 1777, also adopted a female visiting-plan, in this case without the obvious financial incentive. As it transpired, the first two volunteers happened to be wives of governors.[10]

The second decade of the nineteenth century saw a significant rise in the number of institutions, particularly charitable societies and prisons, which admitted women visitors. In addition to Catharine Cappe's success in York, the Leeds Infirmary accepted its first female volunteers in 1816.[11] The Yarmouth workhouse and prison, which will be discussed in some detail later, opened their doors to a woman in 1810 and 1819 respectively. Beginning in about 1810, Susanna Knapp, a niece of John Wesley, spearheaded an attack on Worcester's asylums and prison.[12] Then there was Elizabeth Fry, who first entered Newgate in 1813. Within a few years her associates were visiting prisons and convict hulks around the country. As more and more institutions permitted women to enter their confines, public opinion came around to the view that such schemes were desirable, and this made opposition increasingly difficult. On 20 June, 1820, the General Lying-in Hospital, Lambeth, placed an advertisement in *The Times*, which reflected the change of heart among many of the governors of the nation's public and charitable institutions. It lamented a shortage of funds and invited ladies to visit.

Women visitors usually got their way because they had something to offer. It might be their financial support, their special knowledge, or simply their attentiveness. Domestic know-how and sympathy were particularly needed in orphanages, schools, and in homes for widows, prostitutes, and servants which had large numbers of women or children in their charge. The directors of most of these expanding charities were men, at least in the first half of the century, but there was little in their training which prepared them

[9] Catharine Cappe, *Memoirs of the Life of the late Mrs. Catharine Cappe*, London, 1822, pp. 415–16.

[10] Cappe, *Thoughts*, p. 383.

[11] Cappe, *Thoughts*, p. 384.

[12] Edith Rowley, *Fruits of Righteousness in the Life of Susanna Knapp*, London, 1866, pp. 87–91.

for the internal management of an orphanage or a home for widows. Nor did they speak the same language as the working-class matrons and nurses who staffed their institutions. Thus the assistance of lady visitors became an agreeable proposition.

Initially, such charities often asked the ladies simply to make their observations on the domestic arrangements. The London Infant Asylum, for example, invited women to visit informally and leave their comments in a book left with the matron. But a ladies' visiting committee quickly formed, which gave regular advice on the female side of the administration.[13] Once the women got a foot in the door they extended their duties and their authority. Could fair-minded men resist? As Florence Hill, who worked with juvenile paupers, put it:

> That 'woman's mission' is to tend the young, and nurse the sick, is a proposition constantly urged by the opposite sex, and fully accepted by her own. That 'woman's rights' involve therefore her admission on at least equal terms to the management of institutions whereof the inmates consist chiefly of those two classes, would appear to be a necessary deduction from that proposition.[14]

No less compelling was the view put forward by a leading churchman: 'Female agency, besides being in some respects more efficient, is always less expensive.'[15]

Before long female visitors were powerful figures in the many charitable societies which dealt with causes relating to women and children. It was common for them to inspect the premises several days a week, to oversee the domestic comforts of the inmates, and to supervise the staff. The gentlemen's committee, as it was often called, put few obstacles in their way. The rules of management for many institutions suggest that it often did little more than

[13] *Infant Asylum, for the preserving of the lives, of children of hired wet-nurses, and others*, London, 1799, pp. 32–3.

[14] Florence Hill, *Children of the State: the training of Juvenile Paupers*, London, 1868, p. 264. In 1883 the Moral Reform Union passed a resolution moved by Mrs Charles, a Poor-Law Guardian for Paddington: 'In institutions of which women and children are inmates, and where women are employed as officers, women should, in all cases, have a share in the management.' *The Moral Reform Union. Second Annual Report*, London, 1883, p. 7.

[15] Howson, *Deaconesses*, p. 161.

rubber-stamp decisions taken by the ladies. In the Guardian Society, for example, founded in 1812 to provide a refuge and employment for former prostitutes, the ladies' committee eventually took up the following duties: to direct 'all' the domestic arrangements in the institution; to examine in detail the conduct of all its inmates, 'and advise, encourage, or discharge, as may appear proper or necessary'; to oversee the workshop and to ensure that the matrons carried out their different functions; and last, and perhaps most important, to approve or reject applicants for admission.[16] In short, the ladies' committee, in touch with the day-to-day management, effectively took charge. The gentlemen's committee, happy to lighten its load by delegating authority to what it believed to be expert opinion, accepted the decisions of the ladies. Hundreds of philanthropic societies eventually followed this pattern.[17] By the end of the century it was 'axiomatic . . . that the internal management and discipline' of charitable institutions which maintained women and children should be intrusted to female volunteers.[18]

Power over the internal management of charitable societies gave women fresh ideas about raising money. Committee-men, increasingly aware that they owed a great debt to female subscribers, put up less and less opposition to these schemes. In addition to bazaars and collection boxes, the volunteers experimented with diets and cut down waste in the kitchens. They cleverly turned losses into profits in the workrooms. In the London Female Penitentiary, Pentonville, where subscriptions could not keep up with expenditure, they provided a laundry service and sold needlework produced by the inmates. In the year ending 31 March, 1842, £1,153 came in from these sources;

[16] *Report of the Committee of the Guardian Society for providing temporary asylum, with suitable employment, for Females*, London 1880, pp. 16–17. For another example of 'rules' regarding the duties of a ladies' committee see *An Account of the Adult Orphan Institution*, London, 1831, pp. 36-9.

[17] For a list of institutions commonly run along these lines see [Louisa M. Hubbard], *A Guide to all Institutions Existing for the Benefit of Women and Children*, 5 parts, London, 1878–80.

[18] *Penitentiary Work in the Church of England*, London, 1873, p. 3.

subscriptions came to only £639.[19] Needlework and washing contributed the bulk of the Guardian Society's income by 1880, over four times as much as that raised by subscriptions.[20] Without such innovations, which would not have occurred to most men, many an institution would have been severely restricted in its operations. The perfunctory 'thank you to the women volunteers', common in the annual reports, disguised just how marvellously effective women could be in a time of rising costs.

Administrative and financial innovation was also common in those public institutions which admitted women visitors. Elizabeth Fry and her co-workers devised an elaborate procedure in Newgate in the 1820s.[21] On Monday two women visited the prison, to superintend a reading-class and to distribute books and tracts. Up to four volunteers oversaw the production of baby linen on Tuesday, which was to be sold in the prison shop, run by an agent of the ladies' committee. Profits were to go to the prisoners, half of the sum retained for the day of their discharge (recidivists were to receive smaller shares). Two more volunteers attended on Wednesday, to pass out coarse linen, calico, and flannel to those women unable to work on baby linen. On Thursday, three visitors took charge of the knitting department, reserved for the handful of prisoners incapable of needlework. And on Friday up to five ladies assisted in receiving guests, who were encouraged to buy the work of the prisoners and to inspect the various departments. No one visited on Saturday, which was given over to washing, ironing, and cleaning. Within this framework of cleanliness, godliness, and needlework, the visiting scheme sought to prepare prisoners for a return to society.

The British Society of Ladies for Promoting the Reformation of Female Prisoners, founded in 1821 under the auspices of Mrs Fry, published guide-lines for adoption by local

[19] *The Thirty-Fifth Annual Report of the Committee of the London Female Penitentiary*, London, 1842, p. 27.

[20] *Report of the Committee of the Guardian Society for providing temporary asylum, with suitable employment, for Females*, p. 20.

[21] *Sketch of the Origin and Results of Ladies' Prison Associations, with Hints for the Formation of Local Associations*, London, 1827, pp. 56–8.

associations.[22] These included a set of rules that had been effective in regulating the conduct of female inmates in Newgate. Though established with a prison in mind, these rules could be applied to a wide range of institutions, from orphanages to refuges for aging prostitutes. Begging, swearing, quarrelling, card-playing, novels, plays, and other pernicious books were to be strictly forbidden. As the visitors valued order and morality above all else, work and the Holy Scriptures were to be diversion enough. Monitors, chosen from among the most dutiful prisoners, were empowered to see that their fellow inmates worked and washed with the desired regularity. Any inmate who believed herself mistreated by a monitor had the liberty to complain to the visitors. A policy of distributing rewards several times a year in the form of useful articles oiled the system. It is striking, and revealing, that these rules conformed to those which obtained in evangelical households. The Scripture readings, the warnings against gaming and literature, the love of needles and soap, the discipline and didacticism, were all part of an evangelical upbringing. The women visitors of Newgate moved between the home and the prison with relative ease.

Wherever women had a say in the running of an institution they left their domestic touch. Domesticity was the common experience of women charitable workers, and it should not be undervalued, for the application of household skills to the world outside the home had much to recommend it. Women from large families, and they were common in the nineteenth century, were not ill-equipped to minister and administer in workhouses, orphanages, schools, 'homes', etc. It is unlikely that those they looked after suffered unduly from being treated as part of an extended family. 'I believe that nothing whatever will avail but the large infusion of Home elements into Workhouses, Hospitals, Schools, Orphanages, Lunatic Asylums, Reformatories, and even Prisons', argued Josephine Butler, best known for her campaign against the Contagious

[22] Ibid., pp. 53–6. See also Catherine Frazer, 'The Origin and Progress of "The British Ladies' Society for Promoting the Reformation of Female Prisoners", established by Mrs Fry in 1821', *Transactions of the National Association for the Promotion of Social Science*, vi, 1862, pp. 495–501.

I A working party or Dorcas meeting

II A fancy bazaar at the Wellington Barracks

VIII Sarah Martin visiting Yarmouth Gaol

IX 'Oh! Sir, don't go in'

Diseases Acts.[23] Florence Nightingale, who by her example did so much to stimulate hospital visiting, was another who called on women to deal with both public and charitable institutions as extensions of the home. To her nurses she was specific: 'While you have a Ward, it must be your *home* and its inmates must be your children.'[24] Among the sick the practical realities of life were to renew them and make them more interesting as wives and mothers. It is in this light that we should read her comment: 'The Family? It is too narrow a field for the development of an immortal spirit.'[25]

Institutional work offered women a rich field for their mothering and family instincts. 'It cannot be doubted', wrote Archbishop Tait about his wife Catharine's concern for orphans, 'that the ever-present thought of her own children whom she had lost was an incentive to her care for these destitute little girls'.[26] Josephine Butler's interest in the welfare of children was intimately associated with the death of her daughter Eva. Single women too resolved some of their longings for a family by their work in orphanages, schools, and crèches. Charlotte Sharmon, Baptist, the unmarried daughter of a 'traveller', established an orphanage in West Square, London, in 1867. 'She never ceased to be the mother. . . . There was always delight among them when she came into the home. The babies would hold out their arms to her, and the toddlers would run to her asking for kisses.'[27] Miss Sharmon defined her duties as a 'mother's duties' and obviously received immense satisfaction from her adopted children. Other examples abound. Sister Marie Chatelain, who ran the Catholic House of Charity in Westminster, springs to mind.[28] So does Mary Carpenter, whose

[23] Josephine Butler, ed. *Woman's Work and Woman's Culture*, London, 1869, p. xxxvii.

[24] British Library, Nightingale Papers, ii, Ms included in item 137.

[25] [Florence Nightingale], *Suggestions for Thought to Searchers after Religious Truth*, 3 vols., London, 1860, ii, pp. 388, 393.

[26] The Revd. William Benham, ed. *Catharine and Craufurd Tait*, London, 1879, p. 76.

[27] Williams, *Charlotte Sharmon*, p. 47.

[28] Lady Amabel Kerr, ed. *Sister Chatelain or Forty Years' Work in Westminister*, London, 1900, p. 51.

maternal interest in ragged children has been celebrated.[29] She led an attack on large-scale institutions, for she believed that the orphaned, the sick, and the elderly would reap the greatest benefit from small, cottage homes where they might find a surrogate mother or sister.

Married or single, benevolent women believed that the education they supplied and the rules they enforced would turn their charges into better wives and mothers. The poor were to become imitations of themselves, fit to play their part in managing the nation's domestic affairs. It is not surprising therefore that charitable women transformed many of those who passed through their care into servants, for the skills of a servant were much the same as those of a wife. Domestic service had the great advantage of providing those very desirable home influences. Moreover, training domestics came easily to women volunteers; most of them would not have been on their charitable rounds in the first place if they had not had some practice at it. It was a most agreeable aim to provide employment for the poor while extending their own freedom and influence. Mary Bayly put it more selflessly. The efforts of benevolent women in teaching domestic skills to paupers would lead not only to the elevation of those in need but 'to the amelioration of our whole social system'.[30]

The insatiable demand for reliable servants was an important factor behind the benevolence of late eighteenth and nineteenth-century women, and it played a particularly significant part in their institutional work. Almost every generation of leisured women in England thought itself badly served by domestics; this was never more true than in the nineteenth century. The moral climate and the relative refinement of the age combined to make increasing demands on servants. Furthermore, privileged ladies, anxious to divert themselves outside the home, not least in charitable work, had to depend more and more on the honesty and

[29] Jo Manton, *Mary Carpenter and the Children of the Streets*, London, 1976; Harriet Warm Schupf, 'Single Women and Social Reform in Mid-Nineteenth Century England: The Case of Mary Carpenter', *Victorian Studies*, xvii, March 1974, pp. 301–17.

[30] Bayly, *Ragged Homes and how to mend them*, p. 160.

industry of domestic help. To their dismay, they found their servants very often untrained and unreliable. The 'degeneracy' of domestic servants in the last thirty years, wrote one lady in 1830, 'is universally felt, admitted, and lamented'.[31] Another blamed the chronic shortage of adequate staff on the faulty training that servants received from their betters.[32] To such women the way to an orderly household was to be found in Christian principle and early education. As their own freedom depended on a responsible servant class, they set themselves to work.

In the first half of the nineteenth century they helped to establish charities like the London Society for the Encouragement of Faithful Female Servants and the Bath Servant's Friend Society. But such specialized charities were not alone in showing an interest in the state of the servant market. As we have seen, women around the country were founding or visiting hundreds of orphanages, industrial schools, refuges, homes, deaconess institutions, and aid societies, which catered to the domestic-service industry by teaching outcast women and children the habits of honest labour. The second half of the century saw the most dramatic rise in the number of these institutions. The Reformatory and Refuge Union, which acted as a clearing-house of information for many of these charities, chronicled their growth. By the mid-1870s, the number of its associated societies rose to over 450; and by 1900 it topped 1,000, with 600 of them dealing exclusively with girls.[33] When the Union celebrated its fiftieth anniversary, its members had accommodation for over 60,000 inmates in England and Wales.[34] Not included in these accommodation figures were the discharged prisoner's aid societies, which numbered about 100 by the end of the century. In 1895 they provided services for over 26,000 prisoners, nearly 6,000 of them women.[35]

[31] A Country Lady, *Females of the Present Day, considered as to their influence on Society*, London, 1831, p. 78.

[32] Trimmer, *The Œconomy of Charity*, ii, pp. 3–4.

[33] *Forty-Fourth Annual Report of the Reformatory and Refuge Union*, London, 1900, p. 8.

[34] *Fifty Years' Record of Child-Saving and Reformatory Work (1856–1906)*, London [1906], p. 58.

[35] *P. P., Reports from Commissioners, Eighteenth Report of the Commissioner of Prisons*, 1895, lvi, pp. 124–6.

Complementing the work of the charities associated with the Reformatory and Refuge Union were several large, well-established societies in which women visitors, teachers, and managers helped to promote domestic service. The Ragged School Union had over 600 schools in 1865, by which time it had provided rudimentary instruction for over a quarter of a million children.[36] The Children's Aid Society, founded in 1856, dealt with over 38,000 destitute children in its first fifty years, the girls usually going into service or to schools where they were given a domestic education.[37] Dr Barnardo's Homes,[38] the Waifs and Strays Society,[39] and the rescue homes of the Salvation Army[40] provided a similar service for thousands upon thousands of other girls. Perhaps less celebrated was the work of the Metropolitan Association for Befriending Young Servants, founded by Mrs Nassau Senior, the daughter-in-law of the economist. This institution could boast 25 branch registry offices, 17 associated homes, and over 800 women visitors by the mid-1880s, by which time it was placing over 5,000 pauper girls in domestic employment each year, about 25 per cent of them coming out of the London Poor-Law schools.[41] The training of girls for service was also one of the objects of Miss Louisa Twining's Workhouse Visiting Society, which put thousands of women visitors into the nation's workhouses in the 1860s and 1870s.[42]

Women workhouse visitors, about whom more will be said later, made a considerable impact on the education in

[36] *The Edinburgh Review*, cxxii, 1865, p. 355.

[37] *Fifty Years' Record of Child-Saving and Reformatory Work*, p. 55. See also, *The Children's Aid Society. Its work and its aims* [London], 1938.

[38] Dr Barnardo's Homes trained and placed out over 59,000 children in its first forty years. See *'These Forty Years'. Being the 40th Annual Report of Dr. Barnardo's Homes*, London [1906], p. 37.

[39] Founded in 1881, the Waifs and Strays Society assisted 8,000 pauper children in its first twenty years. See *The First Forty Years. A Chronicle of the Church of England Waifs & Strays Society 1881–1920*, London, 1922, p. 89.

[40] Of the 35,000 females received into the Salvation Army's rescue homes by the end of 1906, over 29,000 were sent to situations or restored to friends. See George R. Sims *et al., Sketches of the Salvation Army Social Work*, London, 1906, p. 95.

[41] *Report of the Metropolitan Association for Befriending Young Servants for 1886*, London, 1887, cover.

[42] See pages 174–81.

Poor-Law schools, not least in turning paupers into servants. The average daily attendance in the workhouse and district schools between 1850 and 1900 was about 30,000; at any one time about 13,000 of these children were girls, who were normally given a rudimentary domestic education which took six or seven months to complete.[43] At thirteen or fourteen the Guardians sent them into service, though in some Unions the demand for the girls was so great that they left as young as ten.[44] It is difficult to be precise about the turnover in the schools, for in many of them there was a constant coming and going of 'casual' children, that is children who still had parents and were in and out of the workhouse as often as six times a year.[45] Probably about 5,000 or 6,000 Union schoolgirls went into service each year after mid-century, most of them directly, some of them assisted by charitable societies.[46] Several thousand of the 'casual' children, many of whom did not attend the Union schools, also joined the servant ranks annually, often with the assistance of registry offices or other charities.

The Poor-Law Unions and the charitable institutions helped to swell the servant class. They effectively produced a body of low-paid domestics, who commonly worked in the homes of artisans and tradesmen.[47] When Mrs Nassau Senior investigated the condition of servant girls for the Local Government Board in 1874, she reported that 'the low rate

[43] For the precise figures each year see the annual *Poor Law Board Reports* and after 1871 the annual *Reports of the Local Government Board*. The numbers began to fall off in the early 1890s, when more and more children were boarded out. The boarded-out girls commonly wound up in service as well. See Ivy Pinchbeck and Margaret Hewitt, *Children in English Society*, 2 vols. London, 1973, ii, pp. 534–55.

[44] *P. P., Reports from Committees, Report from the Select Committee of the House of Lords on Poor Law Relief*, 1888, xv, contains information on workhouse and district-school girls, their training and prospects.

[45] *P. P., Reports from Commissioners, Third Annual Report of the Local Government Board*, 1874, xxv, p. 314.

[46] The Metropolitan Association for Befriending Young Servants found places in 1886 for the great majority of the 1,692 district school girls under its care at the end of 1885. *Report of the Metropolitan Association for Befriending Young Servants for 1886*, cover.

[47] For more on this subject and related issues see my forthcoming article 'Female Philanthropy and Domestic Service in Victorian England', *Bulletin of the Institute of Historical Research.*

of wages given to these girls . . . makes them sought after by many people who, a few years ago, would have done their own housework, whose income does not permit them to keep a superior servant, and who often look on their little servant as a mere drudge'.[48] This trend can be detected from the 1850s onward, when the educational facilities for paupers and prisoners expanded and the number of charities training domestics increased dramatically. To put it conservatively, the total number of pauper and former criminal servants entering the market annually from these sources must have been in the tens of thousands by the 1870s. The vast majority of them were general servants, though the Poor-Law schools and charitable societies sometimes prepared them as cooks. The rapid growth in the female general-servant class after 1850 has frequently been noted. It is perhaps worth quoting the census figures for this, by far the largest class of domestics in the Victorian years:

1851	575,162
1861	644,271
1871	780,040[49]

Just how many of the pauper and former criminal domestics stayed in service is open to question. There is a mass of conflicting evidence in the Poor-Law reports and the accounts of charitable societies. The Hon. Mrs Emmiline Way, who ran the Brockham Home for Pauper Girls near Reigate, testified before the Select Committee on Poor Relief in 1861 that 80 per cent of workhouse girls did not succeed in life.[50] A Kensington Guardian reported to another select committee in 1888 that from his experience 95 per cent of district schoolgirls adjusted to service satisfactorily.[51] The truth

[48] *P.P., Reports from Commissioners, Third Annual Report of the Local Government Board*, 1874, xxv, p. 338. Writing in 1908 W. T. Layton remarked that servants were rarely kept by artisans and the lower middle class before the 1860s. 'Changes in the Wages of Domestic Servants during Fifty Years', *Journal of the Royal Statistical Society*, lxxi, p. 519.

[49] Banks, *Prosperity and Parenthood*, p. 83. The census categories change in 1881 and it is impossible to isolate the general servants from other indoor domestics.

[50] *P.P., Reports from Committees, Fourth Report from the Select Committee on Poor Relief (England)*, 1861, ix, p. 38.

[51] Ibid., *Report from the Select Committee of the House of Lords on Poor Law Relief*, 1888, xv, p. 619.

probably lay somewhere between these estimates, but, as they roughly suggest, there was improvement as the century progressed. This was not accidental, for the Poor-Law schools and charitable societies became increasingly concerned with after-care services. It was largely because of the disastrous lives of former workhouse girls that James Stansfeld, President of the Local Government Board, appointed Mrs Nassau Senior as an inspector of workhouse schools in 1874.

As time passed the government expanded its interest in the work of the voluntary societies which dealt with pauper children. By the mid-1850s the Poor-Law Board 'certified' a growing list of homes, industrial schools, reformatories, and refuges. Certification brought with it government inspection and grants to help pay expenses and place inmates in employment. By the end of the century there were 200 charitable institutions, many of them Catholic, which received government assistance.[52] Most of these were either managed or visited by women, who had not been without influence in persuading the government to legislate. The Act of 1862, for example, which permitted the Poor-Law Board to certify pauper schools, was sometimes called the Emmiline Way Act.[53]

By its legislation Parliament endorsed the view, widely held by women volunteers, that the paupers and delinquents who spent some time in a charitable institution were more likely to succeed in life than those who went directly into service from a workhouse or prison. They often had a built-in advantage, for many charities admitted them because of their right-mindedness or deference, and they were therefore more likely to stay in work and out of trouble. They might also come away from such institutions with a reference.

[52] A list of the charitable societies assisted by government certification is included in *P.P., Reports from Commissioners, Thirty-Fifth Report, for the year 1891, of the Inspector appointed, . . . to visit the Certified Reformatory and Industrial Schools of Great Britain*, 1892, xliii, pp. 44–62.

[53] Louisa Twining, 'The History of Workhouse Reform', *Woman's Mission*, ed. Burdett-Coutts, p. 269. For a list of the Acts which related to the certification of charities see *Fifty Years of Child-Saving and Reformatory Work*, pp. 23–5. And for a discussion of the law relating to delinquent and pauper children see Pinchbeck and Hewitt, *Children in English Society*.

Life without one could be very hazardous, for unlike men who laboured out of doors, female servants found it difficult to improve their situations without a reference. Some of them were, of course, treated kindly, but others were soon broken down. Sometimes they ran away, either to relatives, to brothels, or to North America. Others returned to prison or to the workhouse, or to the charitable society whence they came. Yet a vast number, perhaps the majority, settled down to lives of humble labour, with many moving on to more lucrative positions, or to marriage and the care of their own households. For them it was a hard life, but it could have been worse. Some of the girls came full circle and contributed to the various societies for the rescue of paupers and delinquents.[54] And at least one former servant, Eliza Plomer, became a noted philanthropic worker in the Cripples' Home in Marylebone Road.[55]

The promotion of domestic service was no doubt very desirable to charitable women, but it was less a cause of their work than an effect. Given the class structure of nineteenth-century English society; given the status and power of women in this society; and given the often woeful condition of those trained or rehabilitated, domestic service was one of the only employments that suggested itself. As Louisa Twining said to the Select Committee on Poor Relief in 1861 when asked about the prospects of pauper girls: 'service . . . is the one occupation they can follow in life'.[56] For all its drawbacks, service was work capable of raising the indigent to the level of economic respectability. The women visitors and managers suffered from law and custom too, and this severely restricted what they could do for their charges. They could do little more than prepare them for jobs which market forces and social custom dictated. Moreover, they could only give them the benefits of their own limited knowledge and experience; to a large extent this meant a training in

[54] *A Few Words to Servants about the Church Penitentiary Association*, [1854], pp. 1-3.

[55] *The Reformatory and Refuge Journal*, March 1864, pp. 46-7.

[56] *P. P., Reports from Committees, Fourth Report from the Select Committee on Poor Relief (England)*, 1861, ix, p. 14. See also *Penitentiary Work in the Church of England*, p. 62.

household management. In their enthusiasm to turn paupers and criminals into domestics the charitable women may be accused of self-interest, but this accusation must be tempered by the realization that they were powerless to offer them very much else. In the 1870s, when the issue of 'surplus women' was raging, an 'Office of Lady Helps', headed by Rose Mary Crawshay, opened in London to place middle-class women in service as cooks and maids.[57]

Despite its importance, the training of servants did not take up an institutional visitor's entire day. The time devoted to such work varied considerably from institution to institution. Some inmates already had a skill and looked forward to returning to a job upon release. Others were beyond training, or for that matter, hope. Mrs Fry's assistants in Newgate spent much of their time calming women about to depart for America or the gallows.[58] In addition to their administrative duties, women in charitable institutions might distribute clothing, flowers, or other gifts raised by subscription, write letters, preside over tea and entertainments, or simply carry on a conversation. 'If I had something from the hand of a woman I would get well', remarked a boy to a volunteer in the British hospital in far-off Smyrna in 1855.[59] His comment summed up what the public came to expect from benevolent women—their soft, consoling touch, the blessing of a kindred spirit in surroundings stark and unfamiliar.

The reality was not always so consoling. Inspired by religion, charitable women were not easily deterred in their wish to share it; and so in the nation's institutions, as in its homes, the age-old battle with sin and vice was fought over and over again. The inmates were often unreceptive. One girl, who must have represented many others, said that she did not care to please anyone who could punish her and disliked particularly those visitors who wished her to please

[57] Rose Mary Crawshay, *Domestic Service for Gentlewomen: A Record of Experience and Success*, London, 1874. The experiment was less successful than this title suggests.

[58] Lionel W. Fox, *The English Prison and Borstal Systems*, London, 1952, p. 30.

[59] A Lady [Martha Nicol], *Ismeer; or Smyrna and its British Hospital in 1855*, London, 1856, p. 38.

them before pleasing God.[60] One prostitute, who was locked up and had her hair cut, complained of being hardened by her experience in a penitentiary: 'I was so miserable always thinking about my sins, I couldn't stand it.'[61] She returned to the streets. Despite the set-backs and the embarrassments, the volunteers assaulted their captive subjects with tracts, prayer meetings, and Bible readings, and hammered home the ideal of asceticism and the prospect of eternity. This indoctrination had one overriding aim—to break down deceit and pride and replace it with guilt:

> Bowed with a sense of sin, I faint
> Beneath the complicated load;
> Father, attend my deep complaint;
> I am Thy creature—thou my God.
>
> Though I have broke Thy righteous law,
> Yet with me let Thy Spirit stay;
> Thyself from me do not withdraw,
> Nor take my spark of hope away.[62]

Few things were more disturbing to women visitors than an inmate's unwillingness to confess his or her sin.[63] As with their own children, they believed that severity was needed to produce good effects, to pluck good from evil. Severity went hand in glove with their conception of love. By instilling a sense of shame, or criminality, in their charges, they prepared the ground for conversion and reformation. A defiant inmate was a threat to the security and ordered business of an institution; a contrite inmate, if God willed it, might be redeemed and made productive:

> Such is the fate of guilt, to make slaves tools,
> And then to make them masters.[64]

[60] Mrs Susanna Meredith, *A Book about Criminals*, London, 1881, pp. 200–1.

[61] Ellice Hopkins, *Notes on Penitentiary Work*, London, 1879, p. 5.

[62] Quoted from A. M. Toplady in Meredith, *A Book about Criminals*, p. 145.

[63] Louisa Twining, *Readings for Visitors to Workhouses and Hospitals*, London [1865], is a good example of a manual for lady visitors, who were to bring their charges 'to a sense of their sin, and lead them from the error of their ways', p. x.

[64] Quoted in Meredith, *A Book about Criminals*, p. 190.

Through religious instruction the women volunteers hoped to return the nation's pariahs back to society, cleansed and humble. And as 'physicians of the soul',[65] as interpreters of the sacred mysteries, they added greatly to their own power over others and thereby to their self-regard. Perhaps the writer Eliza Lynn Linton had institutional visitors in mind when she remarked that 'there is scarcely a woman who does not think herself a minor St. Peter with the keys of heaven and hell at her girdle; and the more conscientious she is, the narrower the door she unlocks'.[66]

Most volunteers were probably not as severe and self-serving as this analysis suggests. But they had to be reminded that their task was to relieve as well as to subdue. 'Acts of charity and kindness should never be intermixed with the wormwood and the gall' cautioned a Quaker lady.[67] 'Let us, above all, get rid of this beam of self-righteousness out of our own eye', warned Ellice Hopkins, 'remembering that all the pure and blessed moral influences of our life were not given us to enable us to sit in harsh judgement on those *who have never had them*'.[68] She added that society's outcasts rarely had the advantage of a happy home and a mother's love and that terrible circumstances accounted for most of their misfortunes. Such misfortunes could only be countered by compassion and understanding, by a Christian training with a human face. It was not easy to get women and children who had led idle and abandoned lives to adjust to the methodical labour and loss of liberty which their confinement in public and charitable institutions entailed; and many a soul was lost for the want of sympathy and gaiety. 'Does it never occur to us that the horror our Homes inspire in the mass of these poor sinful girls', argued Miss Hopkins, '. . . may be owing to our making the good life so insupportably dull and monotonous'.[69] Both to relieve and to redeem their charges was a considerable challenge; the former required

[65] A Visitor, *Mornings at the Union*, London, 1858, p. 8.

[66] E. Lynn Linton, *Ourselves. A series of essays on Women*, London and New York, 1870, pp. 7–8.

[67] Margaret Woods' manuscript Journal, iv, 1802–07, Box 0, f. 168, Friends Library, London.

[68] Hopkins, *Notes on Penitentiary Work*, p. 36.

[69] Ibid., p. 25.

a lightness of touch, the latter a seriousness of purpose. Too often a fixation with sin got the better of the charitable woman's reputation for forgiveness.

Many women visitors sought to replace the 'barracks' atmosphere in charitable institutions with what came to be called the 'family system', and they rarely relaxed the religious indoctrination.[70] It was predictably strong in orphanages and schools, but it was a feature of virtually every institution. The Reformatory and Refuge Union, speaking for a thousand charities, proudly announced its chief aim: 'to reclaim and elevate the neglected and criminal classes, by educating them in fear of God and in the knowledge of the Holy Scriptures'.[71] Nor was there any backsliding in workhouse wards and prison cells, where carefully chosen texts opened meetings for prayer: 'Every scripture inspired of God is also profitable for teaching, for reproof, for correction, for instruction which is in righteousness.' (2 Tim. 3: 16–17.) In hospitals, infirmaries, and medical missions volunteers dispensed large doses of Scripture to patients in the waiting-rooms, with special emphasis, we are told, on the death of Jesus.[72] Inside the sick wards their solicitude increased. Here, largely excluded from clinical matters, they were free to concentrate on saving souls and smaller mercies. Their vigilance at the bedside was phenomenal; and it showed itself with particular force during times of unusual distress.

The London Hospital during the cholera epidemic of 1866 provides an example. It may be taken as typical of the visitors' single-mindedness at such times. Doctors, nurses, the Hospital Chaplain, local clergy, members of the Hospital Committee, Scripture readers, and women visitors moved through the wards as the disease gathered up its victims. The ratio of religious practitioners to medical ones was high, a comment on expectations for the patients and the concern of the Hospital authorities for their spiritual welfare. Among

[70] Mrs [Mary] Bayly, *The Life and Letters of Mrs. Sewell*, London, 1889, p. 194.

[71] *Fifty Years' Record of Child-Saving and Reformatory Work*, p. 63.

[72] Margaret Maria Gordon, *The Double Cure; or, What is a Medical Mission?*, London, 1869, p. 17. See also Louisa Clayton, *The London Medical Mission. What is it Doing?*, London, 1873.

the visitors were the 'three Catherines': Catherine Gladstone, whose husband was to become Prime Minister, Catharine Tait, whose husband was to become Archbishop of Canterbury, and Catherine Marsh, the unmarried daughter of a vicar from Colchester. Miss Marsh, a writer of devotional books, left a vivid description of her labours. She went from bed to bed, day after day, for the four months that the cholera persisted. Many of the patients were beyond hope. And as most had not the strength to listen for many minutes, she restricted her message to variations on a single theme. 'Behold the Lamb of God which taketh away the sin of the world', or 'believe on the Lord Jesus Christ and thou shalt be saved'. She repeated these lines over and over again as the cases came in and out of the wards. A woman and her four children were carried in. The mother watched the death of one son and a daughter in the first hour, 'and a sweet young girl of thirteen lay apparently dying at her side, whilst a pale baby of five months old lay in her arms. After a few words of sympathy, she replied in a calm and solemn tone—A heavy calamity has come upon me; but I cannot doubt the kindness of the One who sent it. He spared not his own son but gave him up for us all.'[73]

Another poor woman entered the ward, too ill to talk, who listened attentively to Miss Marsh as she spoke of Jesus. The next day the woman's husband sat by the bedside, weeping. Miss Marsh returned and after a while the woman awoke, recognized her visitors, and said: 'Oh, it's the lady. I'm so glad to see you, dear. I wanted to tell you that I have long known that Saviour you have spoken to me about. He has been precious to me for years, and He will not give me up now.' Miss Marsh intoned:

> How sweet the Name of Jesus sounds
> In a believer's ear;
> It soothes his sorrows, heals his wounds,
> And drives away his fear.[74]

Within the hour another patient lay in the woman's bed.

[73] [C. M. Marsh], *Death and Life*, London, 1867, pp. 33–4, 42–3. See also L. E. O'Rorke, *The Life and Friendships of Catherine Marsh*, London, 1917, pp. 230–1.

[74] [Marsh], *Death and Life*, pp. 46–8.

Christianity may have made it more difficult to live, but if we are to believe the many nineteenth-century witnesses to the Faith, it made it easier to die. Yet the cholera wards were, as a lady nurse told Miss Marsh, an unhappy place to discover this truth: 'None should leave it to a last illness to seek peace with God, for it is as much as they can bear then, TO HAVE TO DIE.'[75] This worry did not stop the visitors from pursuing souls to the bitter-sweet end. They knew, perhaps better than anyone, that Christianity, born with a death, fed on the dying. This was part of the power of their faith. In the cholera wards and the countless sick beds, it gave an answer, however melancholy, to the problem of evil, and death. Could Christianity survive the separation of the dying from the living? The question did not arise in the nineteenth century. The death-bed scene, with its ritual women visitors, reminiscent of that most famous scene at the foot of the Cross, is one of the most enduring images of that century's life and literature.

Not all of those in the bed of sickness passed on to their reward. If they had there would have been far less work for the charitable. Minor ailments often led to serious consequences in the nineteenth century; when cholera or some other infectious disease broke out, the complications were awesome. In their wake were not only the dead but the disabled, the orphaned, the widowed, and the homeless. This human residue posed the most compelling problems for philanthropists. When the cholera subsided in 1866, for example, the 'three Catherines' could not wash their hands of it. Mrs Gladstone, who had promised dying mothers that she would look after the surviving children, lived up to her word: 'She carried off the babies rolled up in blankets.' Some went to an orphanage at Hawarden, while others entered her Home at Woodford, Essex, which she founded as a direct result of the epidemic.[76] Catharine Tait placed many of the girl survivors in a house hired in Fulham and soon established St. Peter's Orphanage, which she visited almost daily when she was in London.[77] For her part, Catherine

[75] Ibid., p. 15.

[76] Mary Drew, *Catherine Gladstone*, London, 1919, pp. 247–8.

[77] Benham, *Catharine and Craufurd Tait*, p. 75.

Marsh opened a convalescent home in Brighton.[78] Epidemics in nineteenth-century England invariably left charities as their monuments, many of them established and supported by women.

Resolution and compassion were the great virtues of female volunteers and times of unusual distress brought them to the fore. It was particularly important that women of social position be on call in such emergencies, for where they trod others were likely to follow. Women like Mrs Gladstone, Mrs Tait, and Miss Marsh, who were all friends, helped to set the tone in such calamitous times. But they were not inactive in happier days. Mrs Gladstone, for example, was a model of charitable enterprise. The London Hospital and the Home which bore her name played only a small part in her philanthropic life. She also worked for the Institution for the Blind, a Home in Notting Hill, and a refuge for prostitutes in Paddington. In hard winters she set up soup kitchens in St. George's in the East, and during the Lancashire cotton famine of 1861 she organized relief. She supported schools and nursing schemes, and not least of all, brought others into her plans. A day in her life at Hawarden, when she was eighty, is related by her daughter:

> She had been to early church, nearly a mile uphill, walking both ways; she had read family prayers at home; she was at her breakfast when word came that a nurse looking after typhoid patients . . . had sickened with fever. Not a moment did she lose, and in her pony carriage she hurried off to Queen's Ferry, where the nurse was lodging. Having made full arrangements, she . . . whipped the nurse off by train to Chester. Arrived there, she supported the patient up and down the long stairs at the railway station, carrying her bag and parcels in a fly with her all round Chester, in vain seeking admittance. At length, partly cajoling, partly scolding, she persuaded the Infirmary authorities to take her in; and, having seen her comfortably tucked up, she returned to the station, with a sandwich from the matron, and reached home about four o'clock.

After tea:

> Flew across to the Orphanage and Home of Rest to charter an audience from the inmates, among whom she placed the Prime Minister, wheedled out of his Temple of Peace; gathered the children round her in the

[78] O'Rorke, *The Life and Friendships of Catherine Marsh*, p. 321.

green-room, and, after a rapid coaching and coaxing, put them through their paces—taking a prominent part herself—and somehow or other contrived to get them through fairly creditably. . . . Afterwards she presided at their tea-party, finishing up by playing spirited dances for them till it was time for them to leave.[79]

Dancing or praying, attending the living or consoling the dying, women like Catherine Gladstone rarely relaxed. Pressures from below as well as above made certain of that. The discipline imposed by faith made fearful demands on the individual which society reinforced. Appeals came from all directions. There were charities to be visited, bazaars to be opened, meetings to be attended, and countless other charitable duties that consumed time and energy. Once the newspapers and magazines took up a woman as worthy of coverage, the demands increased, for charities required public figures to give them publicity and respectability. It would be misleading to imply that such attention disturbed women of benevolent instinct. 'With ever ready hand and heart', to use a contemporary phrase, they took few holidays from God's work. Indeed, on holiday they could be seen passing out tracts and trying to convert other travellers in hotels, on beaches, or country walks.[80] A refusal to accept some charitable task tended to result in an agony of soul. Most tasks were accepted willingly, if not eagerly. In old age, Lady Victoria Buxton, another tireless champion of the evangelical cause, scolded her children for their lack of enterprise: 'But couldn't you get in some little bit of practical work? Couldn't you visit a hospital for instance? Oh how I should love to do it.'[81]

Down the social ladder from Lady Buxton and Mrs Gladstone, the spiritual discipline was no less rigid. Christianity as interpreted by the likes of Wilberforce, Hannah More, and Lord Shaftesbury, instructed the lower orders to be 'diligent, humble, patient; reminding them that their more lowly path

[79] Drew, *Catherine Gladstone*, pp. 244–6. In turn the daughter Mary grew up with 'a great parochial sense' and worked actively as a girl as a Sunday school teacher and district visitor. See *Mary Gladstone (Mrs. Drew). Her Diaries and Letters*, ed. Lucy Masterman, London, 1930, pp. 23–4.

[80] Reaney, *Our Daughters: their lives here and hereafter*, pp. 237–8.

[81] G. W. E. Russell, *Lady Victoria Buxton*, London, 1919, p. 212.

has been allotted to them by the hand of God'.[82] But God was not so class-conscious in respect of individual accountability. The poor must accept their lowly station, but this did not acquit them of their social obligations, which had to be fulfilled if they were to achieve 'the same heavenly inheritance' as the rich.[83] Nineteenth-century Christianity tended to give the poor duties without privilege and stressed the importance of good works as a sign of ultimate mobility. Religious propaganda, dating at least to Hannah More's *Cheap Repository Tracts*, encouraged the poor to walk humbly and to act charitably. Part of this propagandist campaign pointed to women of inferior station as worthy examples of charitable service. Publishers like the Religious Tract Society pumped out an unending stream of news on the activities of English philanthropists, and they were anxious to show that no one class or sex had a monopoly of good works.

The story of Sarah Martin, the workhouse and prison visitor of Great Yarmouth, is probably the best example of a humble philanthropist taken up by the religious establishment in the nineteenth century. Women's magazines, penny tracts, and the lofty *Edinburgh Review* celebrated her work; consequently she is one of the few working women to appear in the *Dictionary of National Biography*.[84] Let us not jump to the conclusion that she was the dupe of insidious religious propaganda. If she was a dupe she was a willing one. Her social superiors did use her story to serve their own interests; but accepting that Miss Martin's work was valuable, they did so in the interest of a wider cross-section of society as well. A spirit of service and self-sacrifice runs through the pages written by her own hand, and this was what her admirers sought to extend by retelling her story. They knew, as Dicken's put it, that 'it is very much harder for the poor to be virtuous than it is for the rich; and the good that is in them, shines the brighter for it'.[85] Miss Martin represents

[82] Wilberforce, *A Practical View*, pp. 485–6.

[83] Ibid.

[84] The Religious Tract Society published a life of Miss Martin in 1860 and one in German in 1862. More recently, see Constance Wakeford, *The Prisoners' Friends: John Howard, Elizabeth Fry, and Sarah Martin*, London [1917].

[85] Charles Dickens, *American Notes*, 2 vols., London, 1842, ii, p. 204.

those countless working-class philanthropists who, in the interests of social harmony, respectability, and improved conditions of life, subscribed to charities, set up visiting and temperance societies, taught in Sunday schools, ran bazaars, and joined in other mercies for their fellow men. She represents the poor of Church and King, of deference and quiet progress; she is thus an antidote to those who, for whatever reason, wound up in the workhouse and prison she sought to reform.

But who was Sarah Martin and what does her life tell us about institutional visiting? She was born in 1791 in Caister, a village three miles from Great Yarmouth, the only child of a small tradesman and his wife. Orphaned at an early age, she moved into a cottage with her grandmother, a glove-maker. She may have attended a dame school, but she was largely self-educated. The religious instruction given to her by her grandmother appears to have made little impression on her as a child. She admitted in later life to having had an 'indescribable aversion' to the Bible in her youth. At fourteen her grandmother put her in the dressmaking trade and after about a year she worked for several local families making and repairing garments. Plain and bookish, there was little else to distinguish her at this time. If she had any male companions, she neglected to mention them in her autobiography. She must have gone through the usual perils of growing up, though in her case the awakening of sex was complicated not by religion but by religious doubt.[86] With few friends and no immediate family except her grandmother, her adolescence must have weighed very heavily on her.

At nineteen her life changed dramatically. During a Sunday outing in the spring of 1810, she attended a sermon out of curiosity at the New Meeting-House in Great Yarmouth. The text was 2 Corinthians 5: 11: 'Knowing therefore the fear of the Lord, we persuade men.' 'It was then', she said in later life, 'that the Spirit of God sent a ray of light upon my guilty soul, slave of Satan, "fast bound in misery and iron" '.[87] The sermon did not result in her immediate

[86] Her poetical work is full of what these days we interpret as sexual metaphor.

[87] *A Brief Sketch of the Life of the Late Miss Sarah Martin*, Yarmouth, 3rd ed. 1845, p. 3.

conversion, but it did force her to examine the state of her soul and the doctrines of the Church. Some months later, in the autumn of 1810, after a searching analysis of the Gospels and some theological writings (from one of her quotations it is probable that she read Wilberforce's *Practical View*) she

> became convinced, not only of the truth of divine revelation, but also that my own crime in having rejected it, embodied guilt capable of every possible manifestation, when not held back by God Himself. By the light of divine majesty, and by His law, I saw myself condemned, and I felt the justice of my condemnation: for not only had I violated that righteous and holy law, but I had added to it contempt of the blessed gospel, and rejection of the Son of God. And yet such was the pity of my God, and such His tenderness to me, that in the immediate disclosure of these my circumstances, He shewed to me, as in the same glance, the Mediator Jesus Christ, my Saviour, and forgiveness through Him.[88]

But even then, after 'looking unto Jesus', the battle with sin continued. It was another year before she felt that 'glorious liberty wherewith Christ had made me free'. Now, she added, 'I wished to give proof of my love, and desired the Lord to open privileges to me of serving my fellow creatures, that happily I might, with the Bible in my hand, point others to those fountains of joy, whence my own so largely flowed.'[89]

After more than a year of anxiety, she had accepted Christ and the responsibility of good works. But by her own admission her charitable work began in the spring of 1810, at the time that she was shaken by 2 Corinthians. It was during this initial phase of her spiritual crisis that she started Sunday school teaching and first entered the Yarmouth Workhouse; and it was then that she first thought of visiting the Yarmouth Gaol, to read the Scriptures to the prisoners. 'I thought much of their condition, and of their sin before God: how they were shut out from the society whose rights they had violated, and how destitute they were of that scriptural instruction, which alone could meet their unhappy circumstances.'[90] But whose soul was more impoverished? Was she too not a sinner, shut out from society? Who was she hoping to convert? Like a great many other charitable

[88] Ibid., p. 4. [89] Ibid., p. 8. [90] Ibid., p. 9.

women, a few of them already noted, her eagerness to serve was touched off by spiritual torment, or by a fear of backsliding, and not by a certainty of grace. This is a clue to the relationship of evangelicalism to benevolence. With their rejection of predestination, typified by the writings of Wesley and Wilberforce, the evangelicals created fresh problems and challenges for themselves. Provisional and conditional, grace required constant nursing.

'A few slight difficulties', which included a hostile turnkey, prevented Miss Martin from visiting Yarmouth Gaol for nine years. But in 1819 she received permission, on her second application, to see a mother who had been locked up for childbeating. 'When I told the woman, who was surprised at the sight of a stranger, the motive of my visit, her guilt, her need of God's mercy, etc., she burst into tears, and thanked me, whilst I read to her the twenty-third chapter of St. Luke.'[91] This passage was apt in more ways than she suspected, for it was the beginning of her own trial in the prison, which was to last for twenty-three years. As she quickly discovered, life behind bars was not very agreeable. The prisoners, who frequently tried to escape, were an unsavoury lot, often filthy and syphilitic. For amusement they fought, swatted flies, and gambled with cinders; for punishment they were flogged or thrown in 'the hold'. The corporation, which was responsible for the administration of the gaol, did not help matters by its parsimony. There was no money for soap or education and very little for medical treatment. Nor was there any real classification of the prisoners, the tried and the untried, the young and the old, being thrown together. Discipline barely existed. The keeper and the turnkey appear to have been more interested in lining their pockets than in enforcing discipline. What order existed, according to the prison inspector William John Williams, came from the work and instruction provided by Miss Martin.[92]

Religion was the reason why she first entered the prison, but it did not limit her interests once she was inside. The promotion of Christianity was, to be sure, her principle aim in reading the Scriptures and teaching the inmates reading

[91] Ibid.

[92] Ibid., pp. 88–95 gives extracts from the reports of inspector Williams.

and writing. But as Williams reported, she also suggested and oversaw their employments, which included book-binding for the men and needlework for the women; furnished them with temporary lodgings upon release; wrote to their relatives; and paid frequent visits to those who settled locally.[93] Details regarding the prisoners, their crimes, education, and conduct, she meticulously entered in financial accounts, journals, and 'every day books'. As her philanthropic schemes expanded, costs mounted. Thus she had to go hat in hand to her neighbours, which at first caused her some unease, for she feared that those who employed her as a dressmaker might be embarrassed by her association with convicts. But as word of her benevolence spread, support was forthcoming. In 1822 a local lady paid her the equivalent of a day's wage, to enable her to rest. She added the day to her teaching commitment. Interest of £10 to £12 a year on an inheritance from her grandmother also went into her work. In 1836 annual donations from Mrs Fry's British Ladies' Society began; she used the money to set up a fund for discharged male prisoners. Some local contributions went into a clothing club, whereby the inmates made shirts, coats, and babies' clothes and sold them for charity. By this plan she turned an initial investment of £7.7 into £408. In 1841 the Yarmouth Town Council persuaded her to accept a grant of £12 a year, in recognition of her services to the community. By now financially independent, she was free to give up dressmaking and devote her time wholly to prison work, 'the highest elevation of desire and satisfaction that I could contemplate, on this side of heaven'.[94]

One aspect of Miss Martin's work deserves special mention. For want of a clergyman willing to volunteer, she led the prison congregation in Sunday services. From 1819 to 1832 she read only printed sermons, but from 1832 to 1837, having built up her courage, she read her own compositions. Inspector Williams attended one of her services in 1835 and was deeply impressed by her stylish delivery and moral authority.[95] The chaplain to the corporation also considered Miss Martin's sermons wholly admirable, and eventually he

[93] Ibid. [94] Ibid., pp. 11–18, 24, 28. [95] Ibid., pp. 25–6. 88–9.

volunteered to take on one of the services himself. After 1837, when a change in prison governors introduced a new regime in the Gaol, Miss Martin's formal compositions were no longer required. But consider what responsibility she had attained when there was no man willing to take on the job! She showed no rancour at the decision to reduce her role at divine services in her autobiography.

The last sermon written by Miss Martin was from Job 19: 25: 'I know that my redeemer liveth.' She asked that it be read to her prison family by a friend on the Sunday following her death, a wish duly observed in the autumn of 1843.[96] In April of that year her prison work came to an end, for her health deteriorated and she was confined to bed. During these last months she revised her autobiography (hoping thereby to reduce its egotistical character), wrote a volume of religious verse, 'In the Sick Room', and concentrated her mind on the atoning death of Christ. Only a short time before her own death, two women visitors sat and prayed at her bedside. In a letter, written in pencil because she could no longer hold a pen, she thanked them for their attentions and added a wish that her many friends would find departure from this world less painful than it was for her. Despite the suffering, her greatest treasure was 'the value and power of the faith of the everlasting gospel of a once crucified but now glorified saviour. In Him may you live;—in Him, may you depart'.[97] A window in the church of St. Nicholas, Great Yarmouth, erected in her honour, depicts the poor women of Joppa displaying the garments sewn out of love and piety by their dead friend Dorcas.[98] It was a fitting tribute to a dressmaker turned philanthropist.

[96] The sermon was read by the Revd. J. E. Fox and is included at the end of her autobiography. The exact day of her death is unclear; at least three different days are given by her various biographers. Given the internal evidence, October 15, 1843 is the most likely date.

[97] Ms letter, Miss Martin to anonymous correspondent [Mr Turner?], 13 Oct. 1843, included in the British Library copy of *Selections from the Poetical Remains of the late Sarah Martin*, Yarmouth [1872].

[98] [George Mogridge], *Sarah Martin, the Prison-Visitor of Great Yarmouth* [1872], p. 11. By her will Miss Martin left all but £15 of her estate to the British and Foregn Bible Society. This amounted to just under £176. Public Record Office, Prob 11/1990, f. 886., 1843; *Forty-First Report of the British and Foreign Bible Society*, London, 1845, p. 96.

Within a few years of her death, Sarah Martin was widely known among women philanthropists. She would have been pleased but probably embarrassed had she lived to see her name mentioned alongside Mrs Fry's.[99] A comparison of the two women raises some interesting issues. As contemporaries they knew about one another's work. Mrs Fry, as we have seen, gave Miss Martin financial assistance after 1836. Both were tireless champions of religious revival; and both sought to introduce habits of order and decency into prison life. Discipline tempered by kindness perhaps best describes their view of the way in which prisoners should be treated. The differences between the two women are no less striking. Mrs Fry, unlike Miss Martin, was well-connected. She had friends like Thomas Fowell Buxton, who would write to the prime minister on her behalf when she wished to visit a particular institution.[100] Such obvious advantages made her demand more and expect more. She was in the limelight after her prison visiting began in earnest in 1817, and she gave evidence before committees investigating prison conditions. (No committee ever invited Miss Martin to be a witness.) Although Mrs Fry was asked for advice, it was usually ignored. She had, after all, called for separate buildings for women prisoners, female officers in the women's wards, an expanding role for women visitors, and a series of other reforms which she believed would make the lives of all prisoners, male and female, less brutish. She saw few of these reforms enacted in her lifetime. On the issue of the role of women visitors, she saw regression. As one authority remarked, everything that she 'asked of Parliament for her women in 1818 was granted by Parliament—in 1948'.[101] Miss Martin, with a sure sense of the limitations imposed upon her class as well as her sex, asked for less and had fewer regrets.

[99] See, for example, Wakeford, *The Prisoners' Friends: John Howard, Elizabeth Fry, and Sarah Martin.*

[100] Buxton lent his support to Mrs Fry's attempt to get a ladies association started in Millbank Prison in 1836. Taking the opportunity provided by the critical Newgate Inspector's Report, the Chaplain of Millbank Prison, the Revd. Daniel Nihill, wrote an attack upon Mrs Fry's methods and intentions to the Superintending Committee. It is clear that he saw a ladies' association in terms of a challenge to his own competence and authority. Public Record Office HO 20/3.

[101] Fox, *The English Prison and Borstal System*, p. 29.

The Whigs set up a committee to look into the state of the nation's prisons in 1835, in part because of pressure from the evangelical interest. It recommended a policy of government inspection, which was to create a system of uniform treatment in the country. The early reports of the inspectors disclosed a good deal about the activities of Mrs Fry and Miss Martin. Inspector Williams, a frequent visitor to Yarmouth Gaol, admired Miss Martin without reservation and attributed much that was orderly and moral in the prison to her influence. As he discovered, the freedom she enjoyed in the Gaol was to a large extent the result of an indifferent corporation, which was happy to encourage her work as long as it did not cost them very much. Moreover, as a lone and humble female visitor with limited reformist aims, she posed no threat to their authority. She thus had no serious clashes with the corporation over the management of the prison; and though her powers were reduced after the reforms of the mid-1830s, she remained active and appreciated.

The inspectors of Newgate were not so uncritical of Mrs Fry and her assistants. They admitted the contribution made by the ladies to the introduction of moral instruction, but they pointed out several features of their work which were 'inexpedient and injudicious'.[102] They accused the lady visitors of interfering with regulations regarding divine services and for going beyond their brief in directing the classification of prisoners and the appointment of female monitors. They criticized them for bringing in too many visitors and for entering the men's wards (Miss Martin visited the men at Yarmouth Gaol without difficulty). And they singled out the prison shop, overseen by the ladies, as 'productive of much evil'.[103] In short, the inspectors thought Mrs Fry and her associates prone to soft-heartedness and meddling. Following their recommendations, the prison authorities tightened the rules on visiting; placed the religious literature circulated by the ladies under strict control; closed the shop; abolished the female monitors drawn from the ranks of the inmates; and

[102] *P. P., Reports from Commissioners, Reports of the Inspectors of Prisons of Great Britain*, 1836, xxxv, p. 19.

[103] Ibid., p. 20.

imposed a silent system.[104] Mrs Fry's ideal of discipline tempered by compassion was not very fashionable in the climate of the 1830s. To most prison inspectors and governors, and many chaplains, the concept of treating prisoners as part of an extended family was alien and unmanly. Silence and the treadmill were the remedies prescribed for the maladies of prison life in the mid-nineteenth century, and in this context Mrs Fry and her associates were largely irrelevant.

There was then a decline in the activity of women prison visitors in the Victorian years; Mrs Fry's experience is but the most obvious example. This was not because of a shortage of volunteers, rather that there was less and less room for them to manœuvre as the government's power over prison life increased. They did not disappear altogether, for the British Ladies' Society was still active in the 1860s, and it sponsored 'a little band of earnest-thinking, persevering women' who received orders to visit prisons and assist the chaplain and schoolmistress; they were especially noted for the help they gave to inmates upon release.[105] But some of their work inside the cells had been eliminated, or taken over by professional Scripture readers. (These government employees, paid about £50 a year, were usually young women who had formerly been district visitors.)[106] Calls to revive the traditional volunteers, to work with Scripture readers, matrons, and chaplains were not particularly well received. An overriding principle of the prison system under the government inspectorate was uniformity, and it was bound to restrict the work of disciples of Mrs Fry and Miss Martin, who wished to treat prisoners as individuals.

The local authorities had never fully appreciated the value of women volunteers, but the Home Office valued them even less. When it took full control of the prison system in 1877, secrecy was added to uniformity. Government bureaucracy

[104] Ibid., *Second Report of the Inspectors of Prisons*, 1837, xxxii, pp. 203-5.

[105] A Prison Matron [Frederick William Robinson], *Female Life in Prison*, 2 vols., London, 1862, ii, p. 168. Though fiction, this work is full of detailed accurate information. See also *Transactions of the National Association for the Promotion of Social Science*, vi, 1862, pp. 495-501.

[106] A Prison Matron [Frederick William Robinson], *Prison Characters drawn from life, with suggestions for prison government*, 2 vols., London, 1866, ii, pp. 143-5.

brought with it what the Webbs called 'a secret world', in which there was little room for the zeal or imagination of a Sarah Martin or an Elizabeth Fry. As the Webbs remarked, there was no longer any room for the investigations of a Henry Mayhew.[107] Mrs Susanna Meredith,[108] a distinguished philanthropist who devoted years to prison visiting and the cause of discharged prisoners, commented on this closed world before the Gladstone Committee in 1894. Asked to discuss the administration of the nation's prisons she replied, with some bitterness, 'I am not able to speak from knowledge of the interior of prisons, because the Prison Commissioners do not permit one's knowing anything about them.'[109] After 1877 her own visits to Brixton Prison had been so inhibited that she no longer wished to return. Her complaint that the Prison Commissioners did not approve of women volunteers was to the point. Sir Edmund Du Cane, the Chairman of the Prison Commissioners, said as much to the same Committee.[110]

In such a climate there was little hope for the recommendations of the more advanced women reformers. Edmund Du Cane would not have warmed, for example, to the suggestion of the temperance leader Mrs Margaret Bright Lucas that the Home Office should establish a separate department for women's prisons headed by a lady secretary.[111] But in spite of 'masculine officialism', as Mrs Lucas put it, women visitors did play a part in the prison system after 1877. In 1892 there were fifty-four prisons which received female prisoners; of these, twenty-nine admitted women visitors, a number that rose during the

[107] Sidney and Beatrice Webb, *English Prisons under Local Government*, London, 1922, pp. 235-6.

[108] See Mrs Susanna Meredith, *Saved Rahab! An Autobiography*, London, 1881; E. M. Tomkinson, *Sarah Robinson, Agnes Weston, Mrs Meredith*, London, 1887, contains a list of the charities run by Mrs Meredith, pp. 105-6.

[109] *P. P., Reports from Commissioners, Report from the Departmental Committee on Prisons, minutes of evidence*, 1895, lvi. p. 188.

[110] Ibid., p. 374. Du Cane complained that lady visitors did not get on with the staff, though he conceded that they did some useful work in local prisons.

[111] Louisa Twining, *Recollections of Life and Work*, London, 1893, p. 282. For a discussion of other prison reforms recommended by women in the late nineteenth century see *The Fortnightly Review*, lxix, 1898, pp. 790-6; *The Contemporary Review*, lxxiii, 1898, pp. 803-13.

1890s.[112] The Home Office imposed strict limitations on these volunteers; most of their energies had to be channelled into the problems associated with discharged prisoners. Gradually the Prison Commissioners recognized the usefulness of women visitors and extended their services. They continued to maintain control over their access and their numbers, however, and would not allow their recommendations to be published in the annual reports of the Commission. Despite these discouragements, Adeline, Duchess of Bedford, founded the Association of Lady Visitors in 1901, and it slowly took on more and more responsibility.[113] It was the twentieth-century heir to Mrs Fry's Newgate Association founded eighty-four years earlier. But it had less power.

The relative powerlessness of women prison visitors after 1835 and more especially after 1877 had two important effects. Firstly, the special problems of female inmates, who made up about 20 per cent of the prison population in most years, received less attention than they would have done otherwise. Under the 'blind uniformity' of Sir Edmund Du Cane's regime, 'a prisoner was a prisoner and practically nothing else'.[114] Secondly, women visitors had to shift their attention from the internal affairs of the prisons to giving assistance to discharged prisoners. This they did with enthusiasm, and even the Prison Commissioners had to recognize the service they rendered. Here was a lesson, which benevolent women learned painfully and reluctantly: nineteenth-century central government could be a forbidding world. Charitable societies, free from the unbending bureaucracy of the Home Office and relatively open to female enterprises, were more agreeable places in which to ply their talents and industry.

[112] Hinde, in *The British Penal System*, p. 246, footnote, cites Du Cane's testimony before the Gladstone Committee that all prisons had female visitors by 1895. Du Cane did say this, but it was contradicted in the *Report of the Prison Commissioners*, 1898, xlvii, p. 25, which noted that all but eleven prisons had women visitors.

[113] In 1922, for example, the Association was allowed to visit boys as well as women. It was not until 1922 that men visitors, the heirs to John Howard and Thomas Fowell Buxton, entered the Home Office prisons to assist the male inmates. Hinde, *The British Penal System*, pp. 162, 246, footnote.

[114] Webbs, *English Prisons*, p. 204.

How successful were women visitors in workhouses? Priscilla Wakefield and others had called upon women to visit workhouses as early as the 1790s, yet the practice was slow to develop, at least on a large scale. This was in contrast to prison visiting, which Mrs Fry organized nationally, indeed internationally, by the early 1820s. As the 1834 Poor Law pushed many of their charges into workhouses women district visitors applied for entry to look after them. But workhouse masters commonly rejected their overtures, and some threatened to resign rather than give in to their demands.[115] Few of them looked favourably on the prospect of women roaming about the wards exposing abuses and making recommendations that might add to the rates. This is not to say that women were shut out altogether. Sarah Martin's teaching in the Yarmouth Workhouse was not singular; several committees of ladies had been appointed in other instances.

About 1850 there appears to have been a rise in the number of women who managed to get the better of the authorities and insinuate themselves into workhouses. Perhaps this was because the Poor-Law Boards were taking a greater interest in the care and education of pauper children; perhaps it was simply because more and more women showed a determination to visit and the authorities caved in under the pressure. The more they saw the more they wanted to reform. What was needed, wrote Anna Jameson, was women's 'domestic, permanent, and ever-present *influence*, not occasional *inspection*'.[116] Among the most notable visitors who took this view were Mrs Hannah Archer in Wiltshire,[117] Mrs Emma Sheppard in Somerset,[118] Miss Gilpin in Liverpool,

[115] [Louisa Twining], *Workhouses and Women's Work*, London, 1858, p. 37, note; [Louisa Twining], *A Paper on the Condition of Workhouses*, London, 1858, p. 50.

[116] Anna Jameson, *The Communion of Labour*, London, 1855, pp. 102–3; *The North British Review*, xxvi, Feb. 1857, pp. 160–1.

[117] Mrs Archer proposed a visiting plan to assist orphan pauper girls in 1861, by which women of the Church of England would form associations connected to workhouses. See *A Scheme for befriending Orphan Pauper Girls*, London, 1861.

[118] Mrs Sheppard wrote extensively on workhouse reform, see, for example, *Experiences of a Workhouse Visitor*, London, 1857; *Sunshine in the Workhouse*, London, 1858.

Mary Carpenter and Frances Power Cobbe in Bristol,[119] and the Lady Mayoress, Mrs Finnis, and the Hon. Mrs Sidney Herbert in London. News of their work spread through friends, pamphlets, lectures, and the various activities of the National Association for the Promotion of Social Science. By the end of the 1850s, there was enough support to launch a national organization, the Workhouse Visiting Society, which was to become closely associated with the name of its Secretary, Louisa Twining.[120]

Born in London in 1820, Miss Twining was the youngest child of a large middle-class Anglican family noted for its charitable work. It was not until 1847, however, that she began to visit the poor in the vicinity of her childhood home in the Strand. She did so, as she later wrote, 'perhaps owing to heavy losses and trials in our family circle'.[121] Extracts from her diary reveal a growing interest in women's work in poor relief, and in 1853, influenced by a pamphlet on poor-house administration by a lady visitor, she first entered the Strand Union Workhouse. In London and the provinces she met and visited with other female philanthropists and began to write and lecture on workhouse reform for the National Association for the Promotion of Social Science. In 1858 her visiting in the Strand Workhouse took a new turn. Impressed by a life of Sarah Martin which she read, she received permission from the chaplain to give afternoon services and sermons. Several years later she published a book of religious readings for workhouse and hospital visitors, dedicated to her friend Catharine Tait. Her contacts were extensive and included Mary Carpenter, Ellen Ranyard, Frances Power Cobbe, Florence Nightingale, Mrs Nassau Senior, Anna Jameson, who visited the Strand Union Workhouse with her, and Harriet Plumptre, the sister of the Christian Socialist F. D. Maurice, whose lectures on Moral

[119] Frances Power Cobbe also wrote on the subject, see, for example, 'Workhouse Sketches', *Macmillan's Magazine*, iii. 1861, pp. 448–61.

[120] See Kathleen E. McCrone, 'Feminism and Philanthropy in Victorian England: the Case of Louisa Twining', Canadian Historical Association, *Historical Papers*, 1976, pp. 123–39.

[121] Twining, *Recollections of Life and Work*, p. 111. For further information on her family see Louisa Twining, *Supplement to 'some facts in the history of the Twining family'*, Salisbury, 1893.

Philosophy Miss Twining attended at Queen's College.[122] Talented and well-connected, she was an obvious choice for Secretary to the Workhouse Visiting Society, sanctioned in 1858 by the Social Economy Department of the National Association for the Promotion of Social Science, and headed by William Cowper, MP, the former President of the Board of Health.

The Workhouse Visiting Society drew together some of England's most talented social and moral reformers. With a mixed management of men and women, it was not a society of 'bachelor rule', to use a phrase of Emma Sheppard. On the male side F. D. Maurice and Dr William Farr, the statistician, joined William Cowper, Lord Lyttelton, and others. On the female side, Mrs Herbert, Anna Jameson, Mrs Sheppard, Mrs Tait, Lady Shaftesbury, and a host of others joined Miss Twining. All the women on the Board came from the solid middle class or above, though in the visitors' ranks lower middle-class women could be found. Miss Twining was especially anxious that lower middle-class women should participate, not least because they had more time to spare than upper-class ladies, whose lives were subject to many more interruptions.[123] Recruiting volunteers was never a problem, for workhouse visiting was not only well suited to benevolent women with leisure, it was less wearing on the feet than house-to-house visiting. The recruits promised to promote the moral and spiritual well-being of all workhouse inmates, to comfort the sick, instruct the ignorant, and encourage useful occupations.[124] It was agreed that this was work best accomplished by women, though the management added the proviso that 'in their efforts [the women] must claim the support of all right-minded and benevolent men.'[125]

The Society, whose strength radiated from London, proposed to place its agents in every workhouse in the nation; and

[122] Her connections are detailed in Louisa Twining, *Workhouses and Pauperism*, London, 1898.

[123] Twining, *Readings for Visitors to Workhouses and Hospitals*, p. xvl.

[124] *Journal of the Workhouse Visiting Society*, Jan. 1859, p. 3.

[125] Ibid., May 1860, p. 202. About 85 per cent of the Society's subscriptions came from women; see the financial accounts in the *Journal of the Workhouse Visiting Society*.

its tactics were gauged to meet the anticipated opposition of suspicious Guardians, workhouse masters, and their minions. In a sense the Society represented an attack, sometimes explicit, upon these officials, who the visitors judged to be often incompetent and indifferent. 'It is the insolence of its officials, and the insubordination of its inmates, that make the poor-house . . . *a hell upon earth*', remarked Anna Jameson.[126] Hostile masters were a particular problem, and the visitors had to marshal their forces and guard their tempers to get around them. An essential tactic was for potential volunteers to exploit their personal and family connections with Guardians and Union Chaplains; an introductory letter from such authorities was the surest way to break down the prejudice of workhouse masters. It was difficult to gain entry without such support, for though Miss Twining obtained the President of the Poor-Law Board's sanction in writing, it was not sufficient to get visitors past the gates of an individual workhouse if the master was intransigent.

Once the visitors got inside, the Society advised them to watch their step, to attend at convenient hours, and to communicate with the management in writing. They were also cautioned about going over the management's head. In 1866 a London workhouse dismissed several visitors because one lady complained to the Guardians of intolerable conditions suffered by the inmates.[127] Volunteers were told to give way on all minor disagreements and not to offend the views of the masters or the inmates, especially their religious views. For this reason the Society excluded Catholics from its ranks and discouraged the proselytizing of Catholic inmates.[128] Overzealousness could be counter-productive. The Brownlow Hill Workhouse in Liverpool was put out of bounds to women visitors in 1854, when it was discovered that some of them had been spreading undesirable doctrines among the sick.[129]

[126] Jameson, *The Communion of Labour*, p. 102.

[127] Louisa Twining, *Recollections of Workhouse Visiting and Management during twenty-five years*, London, 1880, p. 60, note.

[128] *P. P., Reports from Committees, Fourth Report from the Select Committee on Poor Relief (England)*, 1861, ix, p. 3.

[129] Simey, *Charitable Effort in Liverpool*, pp. 62–3, 68–9. The ban was removed in 1858.

Despite these difficulties and restrictions, the workhouse visitors made headway. In line with the policy of the NAPSS they gathered information on cases and divided their duties so that individual volunteers could specialize in a particular category of inmate. Many worked in the sick wards and pushed for improved workhouse nursing run along the lines established by Agnes Jones in Liverpool. This was a reform dear to Miss Twining's heart, and one in which she was to see some measure of success.[130] Others prepared able-bodied women and children for domestic service and worked with them upon release. The after-care of workhouse girls was a recurring subject at the social science congresses and various schemes were proposed. They dwelled on the need for careful supervision and a close working partnership with Union officials, local schools, and the employers of the girls. With their emphasis on social casework such schemes resembled those coming into prominence for district visitors. Miss Twining proposed a rather more ambitious plan in 1859, whereby pauper girls of an age when they would normally find themselves in the adult wards of the workhouse would be placed in a home where they would be trained for domestic service and then given assistance in emigrating to the colonies.[131] Like so many other female philanthropists, the immediate social reality gave shape to her solutions; she was highly practical without being sentimental.

With their experience in domestic management it was only a matter of time before the women volunteers of the Workhouse Visiting Society made improvements in the internal arrangements of workhouses. Some of their changes might appear to be of little consequence, but they none the less made life in workhouse wards less dreary and less routine. Libraries appeared in many wards for the first time, offering both religious and secular literature. The visitors adorned the walls with pictures (usually scriptural subjects), presented lantern lectures, arranged teas, and distributed toys. There

[130] *P. P., Reports from Committees, Report from the Select Committee of the House of Lords on Poor Law Relief*, 1888, xv, p. 337. See also *The Nineteenth Century*, xxxv, 1894, pp. 491–2.

[131] *Transactions of the National Association for the Promotion of Social Science*, iii, 1859, pp. 699–701.

were special functions at Christmas. A typical Christmas party took place in the Leeds Workhouse in 1863, when the lady visitors provided trees decorated with flags, fruit, and presents and laid on a drum-and-fife band for the amusement of the inmates.[132] Such festivities perhaps moderated the widespread feeling among paupers that they would rather be in a prison than in a workhouse. They certainly moderated the deadening blandness of workhouse life, which had grown up, as Miss Twining noted, because the system was fashioned entirely by men.

Compassion tempered by system and common sense were the guide-lines laid down by the Workhouse Visiting Society. Its rules, combined with the restrictions imposed by the workhouse authorities, disciplined the volunteers and worked to eliminate the ignorance and overzealousness that many men associated with women visitors in state and charitable institutions. The Society was only as good as its visitors, and its success as a pressure group at the Union level was a tribute to their sophistication and hard work. As the years passed, more and more Guardians and Union Chaplains spoke out on behalf of the visitors and consequently more and more workhouse masters opened their gates. By the end of the 1870s, there were thousands of women volunteers at work across the country, and Miss Twining could write that the 'Chinese Wall of prejudice' had broken down.[133] Having accomplished much of its task the Society closed its offices in Waterloo Place in 1878. The *Journal* had stopped publication in 1865, for the Committee believed that it had fulfilled its function of publicizing the cause and feared that constant repetition of material would be futile. There was no shortage of assistance from other charities. In 1864, Mrs Tait established the Ladies' Diocesan Association to help co-ordinate institutional visiting in the metropolis.[134] The Girls' Friendly Society added its weight by devoting a special branch to visit workhouse girls in their schools and homes.[135] On a lighter

[132] *Journal of the Workhouse Visiting Society*, April 1864, p. 196.
[133] Twining, *Recollections of Workhouse Visiting and Management during twenty-five years*, p. 90.
[134] Benham, *Catharine and Craufurd Tait*, pp. 449–50.
[135] See Brian Harrison, 'For Church, Queen, and Family: The Girls' Friendly Society 1874–1920', *Past and Present*, Number 61, Nov. 1973, pp. 107–38.

note, the Countess of Meath, Mary Jane Brabazon, set up a Workhouse Concert Society;[136] and Annie Macpherson's Home of Industry in Spitalfields distributed flowers to several London workhouses.[137]

Women workhouse visitors gradually persuaded the public of their utility and took on increasing responsibility. Praise came not only from Union Chaplains and Guardians, but from writers in the periodical press, Members of Parliament, a Select Committee on Poor Relief, and not least from the paupers themselves.[138] The appointment of the experienced workhouse visitor Mrs Nassau Senior as the first permanent female Inspector of Workhouse Schools for the Local Government Board in 1874 was a tribute to their work. An even higher tribute came from the public the following year when the voters of Kensington elected the first female Poor-Law Guardian, Miss Martha Merrington, who was a Secretary of the Kensington District Committee of the Charity Organisation Society.

Miss Twining had called for women Guardians and Inspectors as early as 1861, when she first appeared before a select committee on poor relief. She later helped to form the Society for Promoting the Return of Women as Poor-Law Guardians and was herself elected as a Guardian in Kensington in 1884, her long experience as a workhouse visitor being the spring-board to office. Like the other women who were by this time local government officials, she now sat on the other side of the fence. But the problems were not so different from those she attacked as a philanthropist. One of them was all too familiar—the hostility of men. It was not until 1879 that women managed to attend a Poor-Law Conference, four years after Miss Merrington took office.[139] Thus Miss Twining was not surprised to

[136] Reginald Brabazon, ed. *The Diaries of Mary Countess of Meath*, 2 vols., London [1928], i, p. 10.

[137] Clara M. S. Lowe, *God's Answers: A Record of Miss Annie Macpherson's Work at the Home of Industry, Spitalfields, London, and in Canada*, London, 1882, p. 118.

[138] For examples see *The Westminster Review*, cl, 1898, pp. 32–46; *Parliamentary Debates*, cccxxxiii, 22 Feb. 1889, p. 119; *P. P., Reports from Committees, Report from the Select Committee of the House of Lords on Poor Law Relief*, 1888, xv, p. x.

[139] Twining, *Workhouses and Pauperism*, pp. 40–1.

discover that the Poor-Law Boards set up barriers to women's work. But she lived to see this 'Chinese Wall' also broken down. In 1893 the Local Government Board issued an order to empower Boards of Guardians to appoint female visiting committees.[140] More significant was the Local Government Act of 1894, which established the right of women, married or unmarried, to be elected as Poor-Law Guardians, Urban or Rural District Councillors, and Parish Councillors.[141]

Women workers in the nation's institutions made great strides forward in the nineteenth century. They did so with the assistance of sympathetic men and despite the opposition of the unsympathetic ones. In the case of prison visiting they took one step forward and two steps back; but in most cases there was steady advance. Yet their movements remained restricted, for their power was insufficient to change the conditions which bred the problems they sought to remedy. Their solutions, therefore, were rarely more than palliatives. A prisoner who had been visited by Elizabeth Fry and who later spoke to Susanna Meredith poignantly expressed the frustrating limits of their power: 'Yes', she said, 'Mrs. Fry came and talked to me, and it did me no good; here I am, and here I'll be now until I die'.[142] Mrs Meredith tried again to assist this despairing woman, and perhaps she made an impression. The point is that she made the attempt, when few, if any, would have cared. Such incidents gave meaning to the belief, so strong among benevolent women, that crime, misery, and despair were inevitable wherever feminine influence was ignored. They did not all naïvely assume that compassion and religion alone were adequate instruments to cure the nation's ills. Many of them sought, from the 1860s onwards, to widen their influence by moving in the direction of local government; and some, Louisa Twining among them, also began to support votes for women.

[140] *P. P., Reports from Commissioners, Twenty-Third Annual Report of the Local Government Board*, 1894, xxxviii, p. xci.

[141] 56 and 57 Vict. c. 73.

[142] Tomkinson, *Sarah Robinson. Agnes Weston. Mrs. Meredith*, p. 89.

VI

In Streets and 'Dens of Vice'

Benevolent women were unwilling to give up anyone as lost, and with a passion for tidying up loose ends, for which females were supposed to be noted, they extended their crusade to individuals who were commonly outside the reach of district and institutional visitors. Among those they wished to purify were vagrants, navvies, soldiers and sailors, and especially prostitutes. The reformers who specialized in these groups usually had experience as district or institutional visitors, and their object was much the same: they sought to withdraw the sick and degraded from their unseemly habits and haunts and put them in touch with healing, Christian influence. 'Seeking and Saving', the title of a journal devoted to the reclamation of prostitutes, neatly sums up these activities. And where better to seek out the sinful and benighted than at their places of employment and in their 'dens of vice'?[1]

Charitable women could be seen at factory gates, dockyard dining-halls, and workingmen's canteens, with needlewomen in sweat shops and with navvies at their encampments.[2] Lady Henry Somerset, President of the British Women's Temperance Association, led prayer meetings among miners at the coal-face.[3] They invaded Army barracks and boarded

[1] For an introduction to the subject of rescue work see Edward C. Trenholme, *Rescue Work*, London, 1927; M. Penelope Hall and Ismene V. Howes, *The Church in Social Work*, London, 1965; Madge Unsworth, *Maiden Tribute*, London, 1949; Edward Bristow, *Vice and Vigilance*, Dublin, 1978. On America see Norris Magnuson, *Salvation in the Slums: Evangelical Social Work, 1865–1920*, Metuchen, NJ, 1977.

[2] Mary Merryweather, *Experience of Factory Life*, London, 1862; *Memoirs of an Unappreciated Charity*, London, 1879, p. 29; George Everard, *The Starry Crown, a Sketch of the Life Work of Harriett E. H. Urmston*, London, 1898, p. 166; Garnett, *Our Navvies: a Dozen Years Ago and To-day*; [C. M. Marsh], *English Hearts and English Hands*, London, 1858.

[3] Jennie Chappell, *Noble Work by Noble Women*, London [1900], p. 46.

ships of the Royal Navy; there was hardly a British ship afloat that Miss Agnes Weston did not visit to preach her message of temperance and godliness.[4] Many specialized in pubs and beerhouses; some of them simply distributing tracts outside; a few taking their stand on that unhappy ground between the publican and the thirsty sinner.[5] Female agents worked the railway stations to warn newcomers of the dangers from procurers;[6] and better late than never, they watched the police courts 'to win the erring back to virtue'.[7] Others preached in the streets, amidst the gambling and dog-fighting;[8] the more timid addressed passers-by from the church or schoolhouse doorstep in the hope of persuading them to come in for tea and sympathy.[9] Some women braved the nation's lodging-houses, where they passed out tracts and pleaded with the residents to attend public worship.[10] Mary Higgs, a regional Secretary of the National Association for Women's Lodging Houses, dressed up as a tramp herself the better to appreciate the problems of lodgers.[11] 'Within the purlieus of the lowest haunts, wherever the wretched and guilty were hidden', ministering women pursued them.[12] And although clergymen cut them in the streets, police arrested them, drunks and profligates assaulted them, and a few were killed in the cause, they persevered: 'from morning till evening, anywhere, everywhere, drawn by the lure of woe, Christ's follower went to rescue and to bless'.[13]

[4] G. F. S. Daniell, *Aldershot: A Record of Mrs. Daniell's Work amongst Soldiers, and its Sequel*, London, 1879; Sarah Robinson, *The Soldier's Friend. A Pioneer's Record*, London, 1913; Agnes Weston, *My Life among the Blue-Jackets*, London, 1909; Sophia G. Wintz, *Our Blue Jackets. Miss Weston's Life and Work among our Sailors*, London, 1890; Elise Sandes, *Enlisted: or My Story*, Cork and London, 1896.

[5] Ellice Hopkins, *An English Woman's Work among Workingmen*, London, 1875, pp. 53-4. The periodicals of the Salvation Army are another good source for women pub visitors.

[6] See the *Reports* of the Travellers' Aid Society.

[7] Edward V. Thomas, *Twenty-Five Years' Labour among the Friendless and Fallen*, London, 1879, p. 80.

[8] Lowe, *God's Answers*, p. 128.

[9] A Lady, *Hints for Lady Workers at Mission Services*, London, n.d., p. 4.

[10] Rowley, *Fruits of Righteousness*, pp. 88-9.

[11] Mary Higgs, *Glimpses into the Abyss*, London, 1906; [Mary Kingsland Higgs], *Mary Higgs of Oldham*, London [1954].

[12] Rowley, *Fruits of Righteousness*, p. 90.

[13] Ibid., pp. 90-1.

Inevitably, the 'lure of woe' drew benevolent women into the streets and brothels, or 'dens', in an attempt to rescue what many deemed to be society's most unhappy creature. Concern for prostitutes, or 'fallen women', was nothing new in the nineteenth century; it was at least as old as Christianity. But prostitution had proved intractable, a state of affairs with which the law had little to do, for simply selling one's body for profit was not a crime. The inequalities of income and an exacting moral code, which heightened passion when it did not subdue it, worked to make prostitution a flourishing business in England, as elsewhere. So too did the late marriage-rate among men and the incapacitating effect of pregnancy on married women. Despite the alarm of social reformers and medical men like William Acton, who drew up a 'black list' of the causes of prostitution,[14] many people simply shrugged their shoulders and averted their eyes. But to others, especially women reformers, this was an irresponsible stance, for they believed prostitution to be disastrous not only to the misguided soul 'suffered to drift away' but to the social structure at large.[15] Shocked by the number of 'fallen women' in the population, estimates for London ranged from 10,000 to 80,000 in the 1850s,[16] they commonly believed that there was 'probably not a single family in England that has not suffered in some way from this national sin'.[17]

This is the key to understanding why charitable women took such an interest in prostitution—it endangered what they believed to be their preserve. As guardians of the home and of the purity of the family they defended 'the very fountains of the national life'.[18] Fornication corrupted those

[14] William Acton, *Prostitution*, ed. Peter Fryer, London, 1968 (reprint of 1870 edition), pp. 117-18.

[15] *The Quarterly Review*, lxxxiii, 1848, pp. 359-60.

[16] *Meliora*, i, 1858, p. 76. The 'Judicial Statistics', drawing on reports from English and Welsh police forces, determined that there were about 30,000 prostitutes known to the police in the late 1850s and early 1860s. In London in 1863 there were 5,581 women known to the police as prostitutes, or 1 in 576 persons. *P. P., Accounts and Papers*, 'Judicial Statistics', 1864, lvii, pp. viii-ix.

[17] *Meliora*, i, 1858, p. 73. This author was rather more worried about sons contracting venereal disease than daughters becoming prostitutes.

[18] Ellice Hopkins, *The Present Moral Crisis: an appeal to women*, London [1886], p. 1.

things they held to be most dear. In breaking marriage and baptismal vows it poisoned family relations. No less serious was the threat that venereal disease posed to the health of innocent wives. Moreover, prostitution represented a direct attack on femininity and chivalric ideals, however antiquated. 'Where should we find that reverence for the female sex', wrote the essayist W. R. Greg,

> that tenderness towards their feelings, that deep devotion of the heart to them, which is the beautiful and purifying part of love? Is it not certain that all of delicate and chivalric which still pervades our sentiments towards women, may be traced to *repressed*, and therefore hallowed and elevated passion? Whence could chivalry of old have arisen, save out of chastity? and what, in these days, can preserve chastity, save some relic of chivalrous devotion?[19]

Impurity struck at the heart of long cherished Christian ideals. And if allowed to go unchallenged it might undermine society's deepest feelings about women and their mission.

But was there no contradiction here? Was it not inconsistent to find women who would not mix with the more doubtful members of their own class, whichever it might be, befriending prostitutes? Many forthright female reformers welcomed drunken whores into their homes but swore 'never to receive . . . any man, whatever his rank may be, who is known to be a profligate'.[20] If there was a contradiction in this it was not one which bothered them very much. When challenged they replied that their attitudes flowed from the same moral imperative; and they quickly added that prostitution affected women more deeply than men. The fact that women suffered most from loss of purity was never far from their minds, and it made many of them angry. No wonder their heart went out to the despairing Magdalene and not to the roué, who might easily have been the cause of some woman's ruin and who, given his advantages, should have set an example of propriety in the first place. Fallen women, whatever their station may have been, made the most fundamental call on the sympathies of other women. There was

[19] *The Westminster Review*, liii, 1850, p. 254.

[20] *P. P., Reports from Commissioners, Report of the Royal Commission upon the Administration and Operation of the Contagious Diseases Acts*, 1871, xix, p. 447.

more than a little truth in Emma Sheppard's remark that 'a woman's hand in its gentle tenderness can alone reach those whom *men* have taught to distrust them'.[21]

Yet some of the keenest supporters of charitable work drew the line at female volunteers visiting brothels and roaming the streets in search of abandoned sinners. Benevolent women sometimes refused to take on such tasks, for they thought them indelicate and unfeminine.[22] And benevolent men sometimes voiced a worry lest wives and mothers be 'soiled by contact with impurity';[23] and fearing that soft-heartedness got in the way of reforming prostitutes, they occasionally went so far as to oppose the admission of women to charitable committees.[24] But the moral wilderness conjured up by dens of vice did not frighten the most determined women. They remembered the injunction that it was a female's duty 'to promote the purity and morality, the welfare and security of her own sex, and especially of the young, the friendless, and the exposed'.[25] Christian faith, they concluded, was not the poorer for being tested; and it would have to be weak indeed to be undermined by contact with such unhappy members of their own sex.

The worries of charitable women became pronounced in the middle years of the century, when prostitution seized the English imagination. Henry Mayhew's investigations into the breeding-grounds of prostitution for the *Morning Chronicle* in late 1849 and 1850 made a powerful impression, and a few of his case-studies of needlewomen and slop-workers who were compelled to resort to the streets to make their livings were reprinted in the *Westminster Review*.[26] Several books attracted further attention to the subject,[27] most

[21] Mrs Emma Sheppard, *An Out-Stretched Hand to the Fallen*, London, [1860], p. 6.

[22] [Dinah Maria Muloch], *A Woman's Thoughts about Women*, London, 1858, p. 295.

[23] *The Quarterly Review*, lxxxiii, 1848, p. 376.

[24] Ibid., p. 375.

[25] *The Female's Friend*, January 1846, p. 2.

[26] *The Westminster Review*, 1850, liii, pp. 245–9.

[27] In addition to Acton's work see A Physician [Gustave Richelot], *The Greatest of our Social Evils*, London, 1857; A. J. B. Parent-Duchatelet, *De la prostitution dans la ville de Paris*, Paris, 1857; James Beard Talbot, *The Miseries of Prostitution*, London, 1844.

notably Acton's *Prostitution* first published in 1857, and in reviewing them the periodical press popularized the daunting though in some cases dubious statistics compiled. In the same years various charities which dealt with fallen women were publishing their own hair-raising accounts and statistics, some of which were to reappear in Mayhew's *London Labour and the London Poor.*[28] The assorted investigations, reports, and the statistics scandalized many people who had not given much thought to the issue before and reinforced the alarm of the many respectable citizens already familiar with the nation's street life after dark.

Socially conscious Victorians took to facts, or what they believed to be facts, as a terrier takes to a rat; and some of them shook themselves into a near frenzy of moral indignation once they got hold of a few facts about prostitution. The revelations of the social investigators pulled on the heart-strings, gave direction to righteous anger, and opened up for reform a relatively untrammelled field of inquiry. It was this fresh perception that prostitution was a growing menace which lay behind the enthusiasm for brothel visiting and its related activities after mid-century. That these activities were seen as part of a holy war on immorality and impurity, justified by biblical authority and Christian practice, made action all the more imperative:

> Give me the power to draw
> This soul from sin and shame,
> From the madding streets, from the din
> and the roar
> Oh! call Thy child by name!
>
> Lord, here am I, send me
> To do Thy will, oh, God!
> To seek the lost and set them free,
> From sin and the devil's rod.[29]

The charitable had, of course, done much before mid-century about the perceived dangers of prostitution.[30] This

[28] Bracebridge Hemyng, 'Prostitution in London', extra volume, p. 210.
[29] From 'On the Streets' by Ellen Pash, *The War Cry*, 10 April 1886, p. 5.
[30] See Michael Ryan, *Prostitution in London*, London, 1839, pp. 88–211.

was the heyday of the Society for the Suppression of Vice after all. Moreover, casual and district visitors had often tried to assist prostitutes when they came into contact with them on their neighbourhood rounds; the reports of the early district-visiting societies frequently contain statistics on fallen women placed in penitentiaries or returned to their family or friends. More directly concerned were the specialized institutions for the reclamation of prostitutes. These female penitentiaries and Lock hospitals, which treated patients with venereal disease, began to appear in the mid-eighteenth century. But there was a phenomenal upsurge in interest in the mid- and late Victorian years. This was reflected in the dramatic growth in rescue societies and Magdalene homes; the latter jumped in number from 60 in 1856 to 308 fifty years later.[31] Much has been said about the running of homes for penitents already, but it is worth remembering that they were remedial institutions and incapable of checking the prodigious problems. There was always room for experimentation.

The most popular innovation of the Victorian years was 'rescue work', which was a systematic attempt to remove prostitutes from their habitual haunts. Volunteers and paid missioners roamed the streets and visited brothels in the hope of persuading fallen women to enter a refuge or a female penitentiary where they might begin a new life. And as with district and institutional visiting, women carried out most of this labour. Some of the societies were managed by men, the Rescue Society in London for example. But the men interested in this form of charitable endeavour (Gladstone was far from unique) were well aware of their dependence on their female colleagues. They knew from embarrassing experience that male rescue workers were at a grave disadvantage on the streets and in the brothels, for their intentions were likely to be misunderstood. If Gladstone's ambivalence in the company of prostitutes was common, there was perhaps good reason for this.[32] One of the leading

[31] *Fifty Years' Record of Child-Saving and Reformatory Work*, p. 57.

[32] *The Gladstone Diaries*, iii, 1840–7, eds. M. R. D. Foot and H. C. G. Matthew, Oxford, 1974, pp. xliv–xlviii. For another example of a male rescue worker whose motives were sometimes mistaken when he 'cruised' the streets see Lieut. John Blackmore, *The London by Moonlight Mission*, London, 1860.

authorities on rescue work, Arthur Brinckman, Chaplain of St. Agnes' Hospital, argued that men ought not to do it at all. As he put it: 'women can, will, and *must* do the greater part of this work amongst women'.[33]

In time women took a larger share in the management of the rescue societies. They were the backbone of the Female Mission to the Fallen, established in 1858 by a fund set up by the Reformatory and Refuge Union. Indeed, this society called itself a 'woman's mission to women'. Using a combination of lady superintendents and paid working-class missioners, it resembled Ellen Ranyard's organization, which, founded a year earlier, probably influenced it. In providing accommodation for mothers with babies it was itself pioneering. By 1906 it had sent 12,500 prostitutes to homes and refuges.[34] Increasingly well organized, women formed a large number of their own rescue societies and associations, many of which were small community-based charities run along the lines of the parish district-visiting societies and frequently attached to them. Mrs Julia Wightman's rescue society in Shrewsbury is but one example of this type.[35] Another variant was the society associated with institutions for the reclamation of prostitutes, typified by Mrs Anna Wilkes' charity 'Rescue Work in Poplar'.[36] So important were women in this branch of philanthropy that it is fair to say that without them this great campaign would never have taken off.

In rescue work, as with other forms of charitable activity, no one class or denomination had a monopoly. As might be expected, middle- and upper-class evangelicals were in evidence, but working-class Bible-women and home missionaries, usually assisting women of the higher classes, were also very active. So too were Catholic sisters and Anglican deaconesses. Mrs Mariquita Tennant of the House of Mercy, Clewer, and

[33] Arthur Brinckman, *Notes on Rescue Work*, London, 1885, pp. 35, 81. See also A. O. Charles, *The Female Mission to the Fallen*, London, 1860, p. 4.

[34] *Fifty Years' Record of Child-Saving and Reformatory Work*, p. 57; Hall and Howes, *The Church in Social Work*, p. 20; Charles, *The Female Mission to the Fallen.*

[35] [Wightman], *Annals of the Rescued.*

[36] *P. P., Reports from Committees, Report from the Select Committee of the House of Lords on the Law relating to the Protection of Young Girls*, 1882, xiii, p. 40.

the redoubtable Sister Dora Pattison of Walsall, whose brother Mark was the Oxford reformer, set an example for other Anglican nuns.[37] Mrs Edmund Kell of Southampton, who was asked to give evidence on rescue work before a parliamentary commission in 1871, gave a similar lead to Unitarian brothel visitors.[38] Nor should we forget the Jewish ladies, who despite the suggestion that prostitution was not a problem among Jews, formed the Jewish Ladies' Society for Preventive and Rescue Work in 1886.[39]

Adding to the variety were the women of the Salvation Army. Led by Catherine Booth and Mrs Bramwell Booth, they were among the most active and successful rescue workers by the mid-1880s. By 1890 there were about 4,500 full-time Army officers at work in Britain,[40] a large percentage of them working-class women, many of whom had been social outcasts themselves. By the end of the century they claimed to rescue as many as 5,000 females a year, including many prostitutes.[41] The periodicals of the Army, *The Deliverer, The War Cry*, and *All the World*, are richly endowed with stories, woodcuts, and photographs of women going out two by two in search of 'lost sheep', of nightly brigades of up to 300 women making raids on Piccadilly and Regent Street, of pitched tents and battles outside notorious public houses, of bands playing 'Home Sweet Home' and cadets singing softly under the lamps as members of their fellowship passed out invitations to meetings and shelters. No other society was in closer touch with the class of woman being assisted; no other society which had men attached to it

[37] The Revd. T. T. Carter, *The First Ten Years of the House of Mercy, Clewer*, London, 1861, p. 6; Millicent Price, *'Inasmuch as . . .' The Story of Sister Dora of Walsall*, London, 1952, pp. 45-6. Sister Dora's faith defies easy categorization. Evangelicals, High Churchmen, and Catholics all claimed her as their own. See Jo Manton, *Sister Dora. The Life of Dorothy Pattison*, London, 1971, p. 269.

[38] *P. P., Report from Commissioners, Report of the Royal Commission upon the Administration and Operation of the Contagious Diseases Acts*, 1871, xix, pp. 601-6. See also *Memorials of the Rev. Edmund Kell, B. A., F. S. A., and Mrs. Kell, of Southampton*, London, 1875.

[39] Battersea, *Reminiscences*, pp. 419-22. William Acton was told that Jews were not much involved in prostitution, for they looked after one another better than Christians. Acton, *Prostitution*, p. 53.

[40] Booth, *In Darkest England and the Way Out*, Appendix I.

[41] *The War Cry*, Dec. 9, 1899, p. 4.

was more conscious of the importance of female volunteers, more egalitarian in approach, or more closely knit by its sense of Christian purpose.[42]

The precise number of rescue workers is unknown. There were certainly fewer of them than there were household or institutional visitors, though it should be remembered that they commonly took on these tasks as well. Be that as it may, there were eventually enough rescue workers to visit all the nation's brothels, which numbered about 7,000 in the 1860s, and all the hundreds of brothel quarters on a regular basis.[43] It is unlikely that any prostitute in the business for long was unaware of their existence. In some districts, London's West End for example, women representing different religious persuasions and different charitable societies competed for the same sinners. Whatever their faith or affiliation, women rescue workers had at least two things in common: a desire to save souls, and fearlessness. As the High-Church Anglican Ellice Hopkins, perhaps the most indefatigable worker in the cause, remarked: 'We are not going to quench this pit of hell in our midst by emptying scent bottles upon it.'[44]

Reclaiming prostitutes was a daunting prospect for charitable women however tough-minded, and they had to prepare themselves for it, both educationally and emotionally. Many of them first became interested in rescue work through their visits to Lock hospitals and asylums, Magdalene wards of ordinary hospitals, or workhouses. Here, as we have seen, they conversed with fallen women with little difficulty, and picked up information and made contact with other reformers. But they were cushioned against some of the harsher aspects of prostitution in such institutions, for the authorities

[42] Women's rescue work in the Salvation Army deserves more attention. For further information see Unsworth, *Maiden Tribute*; Robert Sandall *et al. The History of the Salvation Army*, 6 vols., iii, *Social Reform and Welfare Work*, London, 1955; Catherine Bramwell-Booth, *Catherine Booth: The Story of her Loves*, London, 1970.

[43] *P. P., Accounts and Papers*, 'Judicial Statistics', 1864, lvii, put the number of brothels and houses of ill fame at 7,204 in the returns of 1863. This was something of a drop from the late 1850s. *The Eighteenth Annual Report of the Rescue Society*, London, 1872, p. 32, contains the 'Judicial Statistics' on brothels and prostitutes for the years 1863 to 1870.

[44] Hopkins, *The Present Moral Crisis*, p. 3.

were never far away and after all, they were dealing with women in a relatively helpless condition. As many of them discovered, trying to reform a prostitute in a Magdalene ward or a workhouse was not the same thing as trying to reform one in the midst of her drunken companions in a port or garrison-town brothel. Yet institutional visiting provided something of an apprenticeship for many women who turned their hand to rescue work. If nothing else it gave them a glimpse of the difficulties ahead and taught them the unlikelihood of easy solutions.

More intimidating than institutional visiting was rescue work in the streets. This was the nearest and, experts believed, the next best thing to visiting the brothels themselves. Here the volunteers typically scoured the brothel quarters in pairs, between eight in the evening and one in the morning. Their object was to isolate a prostitute from her companions and tactfully pass her a tract or what was called a 'liability card', which, with a heading like 'come unto me, and I will give you rest', listed local Magdalene homes and refuges.[45] When possible the rescue worker offered the woman tea and a room for the night in her home, or failing that tried to persuade her to go immediately to a refuge or a midnight meeting. Innovation and quick thinking were at a premium. For the novice there was no shortage of advice. 'Stand at the edge of a crowd round a Band, or Punch, in certain thoroughfares, and mark the girl with cheap but studied toilette, cheeks defiled with rouge, and restless, reckless air, and stealing gently near her, whisper, "Do you want a friend to help you?" "Does your life make you happy?" '[46] One line was standard and said much about what rescue workers held most dear: 'Would you like your mother to see how you are going on just now?'[47] Rebukes were common. As one experienced volunteer lamented, the girls 'generally laughed in their faces'.[48]

[45] There is an example of a 'liability card' included in *Penitentiary Work in the Church of England*, p. 135.

[46] *Seeking and Saving*, i, May 1881, p. 7. The most thorough manual for rescue workers is Brinckman, *Notes on Rescue Work*. For advice to rescue workers in the 1920s see Trenholme, *Rescue Work*, pp. 62–70.

[47] *Seeking and Saving*, i, May 1881, p. 7; J. M. J. Fletcher, *Mrs. Wightman of Shrewsbury*, London, 1906, p. 107.

[48] Robinson, *The Soldier's Friend*, p. 92.

Rescue work could be exhilarating and dangerous. A 'Skeleton Army' of rowdies often shadowed volunteers of the Church Army and for amusement pelted them with red ochre, rotten apples, and, with a taste for wry humour, oats.[49] Few rescue workers escaped altogether unscathed. Mrs Butler and some of her female supporters once found themselves trapped by a mob in a smoke-filled barn, 'like a flock of sheep surrounded by wolves'.[50] But the Salvationists probably suffered more than anyone else. Whether it was their determined faith or the noise they made which caused the greater annoyance is difficult to tell. Assaults were common, and in the early years the police did little to prevent them. Tomatoes, rotten eggs, and dead fish (Catherine Booth once had a cod's head tossed in her lap) did little to reduce the Salvationists' hunger for souls. It is likely that the missiles and insults simply strengthened the comradeship in the Army. But things could get out of hand. One girl lost an eye in a fight with procurers; another woman was 'kicked into insensibility' outside a police station in Guildford; and another was so badly beaten that she died a week later. In Shoreham, a female Captain was 'promoted to Glory' by a stone thrown at a prayer meeting. Hundreds of others were attacked, among them many women and children; and in Torquay between 1886 and 1888 forty Salvationists found themselves in jail.[51]

Mingling with whores on midnight streets was not then for the faint of heart, but with a little practice and a thick skin rescue workers gained confidence and brushed aside the rebukes and rotten eggs. Not a few women came to enjoy the confrontations and the rows that frequently ensued. And occasionally they went beyond the call of duty, or as one Manchester campaigner put it: 'We almost

[49] A. E. Reffold, *The Audacity to Live*, London, [1938], pp. 27-33.

[50] Josephine E. Butler, *Personal Reminiscences of a Great Crusade*, London, 1896, pp. 88-92.

[51] General Frederick Coutts, *No Discharge in this War*, London, 1975, pp. 90-1; Adelaide Cox, *Hotchpotch*, London, 1937, p. 59; K. S. Inglis, *Churches and the Working Classes in Victorian England*, London and Toronto, 1963, p. 192. For a personal view of rescue work in the Salvation Army see Mary Dean Morgan, *The Lady with the Other Lamp. The Story of Blanche Read Johnston*, Toronto [1919].

force them [prostitutes] to come with us that we may do them good'.[52] One ploy, condemned by the Church of England, was for the rescue woman to trail behind a male decoy the better to detect her prey.[53] Another was to interrupt suspicious-looking couples, going so far as to sit purposely between men and women on park benches to break up any potential alliance.[54] Mrs Laura Ormiston Chant, of the National Vigilence Association, spoke with pride of one of her experiences to a meeting of social purity workers. She once spotted a man accosting a young girl late at night on the underground and resolved to see her safely home. Overhearing enough to heighten her alarm, she followed them out of the carriage and 'sent the scoundrel flying off as fast as his feet would carry him'. To the girl she said 'you must thank God that He sent me to save you'.[55] She did not report what she said to the man, or whether the girl thanked her for her interference.

Closely associated with rescue work in the streets was the midnight meeting, another of the stratagems which complemented brothel visiting and which helped to train brothel visitors. On the evening of the meeting the promoters distributed invitations to prostitutes, which promised them free refreshments, usually tea, coffee, and cakes.[56] As the guests discovered, their hosts mixed these delights in about equal portions with religion, dispensed by platform speakers and women volunteers sitting at each table. The idea probably originated with the noted philanthropist Theophilus Smith, the Secretary of the Female Aid Society. The first recorded meeting took place in 1850 in a schoolroom on the Waterloo Road; but as only seven prostitutes attended, another decade passed before he repeated the experiment. After much preparation he and his supporters rented the St. James's

[52] Mrs Arthur Weigall, *Seeking and Saving: being the rescue work of the Manchester Mission* [Manchester], 1889, p. 7.

[53] Trenholme, *Rescue Work*, p. 62.

[54] Charles, *The Female Mission to the Fallen*, pp. 8–9.

[55] *Speech of Mrs. Ormiston Chant at the Annual Meeting of the S. P. A., June 13th, 1883*, London [1883], p. 3.

[56] Examples of midnight-meeting cards of invitation are included in [Theophilus Smith], *Statement of the Origin, Proceedings and Results of the Midnight Meetings for the Recovery of Fallen Women*, London [1860], pp. 52–3.

Restaurant in Regent Street for the night of 8 February 1860. Their object was to entice and to subsequently reform the many prostitutes pouring out of the local casinos at midnight. To this end several workers passed out 500 invitations. About 250 women appeared, not knowing what to expect beyond the prospect of tea and crumpets.

'It was a marvellous gathering', wrote Edward Thomas, the Secretary of the London Female Preventive and Reformatory Institution,

> . . . the dresses of many of those present were absurdly gay and extravagant. Fashion's latest freaks were duly observed, and not a few were decked with at least *apparently* costly jewels. The surprise of the large majority was intense, and constituted the first subject of conversation. Some said they thought the whole thing was a hoax; others, that a fast man, with a view to a spree, had brought them together; but when the real object was discovered some laughed at the idea, a few went away, but the majority remained, determined to see the meeting out.[57]

The Revd. Baptist W. Noel provided the evangelical address, which dwelled on the virtues of family life and female character, and the willingness of Christ to forgive the most erring and sullied creature. Other speakers then offered up prayers. 'A large number of the fallen sisterhood buried their faces in their handkerchiefs and sobbed aloud, whilst more than one had to be removed in an almost unconscious condition from the room.'[58] The meeting ended at about 3 a.m., following an announcement that anyone wishing to forsake a life of sin would be welcome in a nearby home. To the surprise of the police, several girls followed Mrs Edward Thomas through the cold, nearly deserted streets to her refuge.[59] The midnight-meeting movement was underway, and it spread rapidly to other parts of London and the provinces. Within a year twenty meetings had been held, attended by 4,000 prostitutes.[60] In

[57] Thomas, *Twenty-Five Years' Labour among the Friendless and Fallen*, p. 77. For another account of this meeting see *The Magdalen's Friend*, i, April 1860, pp. 27–30.

[58] [Smith], *Statement of the Origin, Proceedings and Results of the Midnight Meetings for the Recovery of Fallen Women*, p. 10.

[59] Thomas, *Twenty-Five Years' Labour among the Friendless and Fallen*, p. 78.

[60] *The First Annual Report of the Midnight Meeting Movement for the Recovery of Fallen Women*, London, 1861, p. 45.

one six-month period in the mid-1880s in Manchester, Mrs Arthur Weigall, Secretary of the City Mission, ran fifty meetings.[61] Such events remained popular right through the rest of the nineteenth century.

We should be wary of the statistics published on the number of prostitutes reformed by the midnight-meeting movement.[62] Not all of the meetings were productive and sometimes they were fiascos. In 1864 Miss Sarah Robinson, the 'soldier's friend', experimented with a midnight meeting in Aldershot. 'We got the tea ready for 10 o'clock', she recalled,

> . . . About 30 girls came. Nearly all were the worse for drink, and it took an hour's hard tea drinking to get them sober and fit to listen. . . . They kept slipping out, on one pretence and another, till only a dozen remained; and one of these went into a fit, and made a long interruption, and another, also, became faint and hysterical, and the Bible women were wholly occupied with smelling salts and cold water.[63]

Miss Robinson expected to get seven girls into a refuge from the night's work, but no one appeared the following day. The problem of drink was not easy to overcome at midnight meetings, so a breakfast variation developed, which was not so likely to be interrupted by alcohol and smelling-salts. In the morning meetings 'everything is real about them, no illusion, no self-deception, no excitement, but real misery, pain, remorse'.[64] Yet such activities did not satisfy the most tenacious rescue workers. Having tried the various expedients, Miss Robinson and many others concluded that there was no substitute for visiting the brothels themselves. As Ellice Hopkins put it, 'Christian visitation of dens by *women* in the daytime, is a far better agency than midnight meetings, with addresses by *men*.'[65]

Brothel visiting was the charitable woman's ultimate challenge, for in her imagination it was not unlike a journey

[61] Weigall, *Seeking and Saving*, p. 23.

[62] Statistics can be found in the annual reports of the Midnight Meeting Movement for the Recovery of Fallen Women beginning in 1861. See also [Smith], *Statement of the Origin, Proceedings and Results of the Midnight Meetings for the Recovery of Fallen Women*, pp. 18–37.

[63] Quoted from Sarah Robinson's Journal in [Ellice Hopkins], *The Visitation of Dens. An Appeal to the Women of England*, London, 1874, p. 125.

[64] Ibid., p. 127.

[65] Ibid., p. 128.

into hell itself. Miss Robinson's initial foray into the row of brothels in Aldershot, which took the name of the adjoining pub, the Shamrock, was not untypical:

> I was very *foolish* about going into the girls' rooms. I walked up and down praying for strength, and feeling wretchedly incapable, and without self-control. The people even in the low lodging-houses advised me not to venture, as I should certainly be insulted. At last I dashed into it, and found it really not so very difficult. Most of the girls were not up; nearly all, after their first surprise, received me kindly. One poor creature, Lizzie, I found in tears; she is ill, and fears she shall die. I talked and prayed with her; she clung to me as if I could save her, and I promised to come again. Nearly 100 girls live in this *Shamrock*, 2 in each room, 'chums', as they are called; girls of all grades, some shockingly diseased. . . . I could not write down, I cannot even bear to think of the horrible things I saw and heard; but I only met with unkindness in one place, where the women who were ironing would have burnt me with the iron if I had stayed.[66]

She was violently sick when she got home; but before long she had a couple of assistants, one of them a navvy, praying and singing outside as she 'dived' into the dens. Up to 500 prostitutes may have heard her each day. And she was reported to have brought five to ten girls a week to a refuge.[67]

Inspired by Miss Robinson's example, Ellice Hopkins began to visit the dens of Brighton with two female friends. The well-educated daughter of a Cambridge mathematician, she had had experience as a Sunday school teacher, a missionary to navvies, and a pub visitor; she had been cut in the streets by clergymen and accosted by drunks.[68] Yet her first experience as a brothel visitor came as a considerable shock. It was 'like dashing myself against a dead wall' she recalled. Exasperated by her reception, she exclaimed to several girls: 'I can't stand this . . . you must kneel down and pray with me.' They did, but before the door could be closed behind her she heard 'shrieks of horrid laughter' and 'fragments of indecent jests'. Shattered by the experience she had

[66] From Miss Robinson's Journal, quoted in [Ellice Hopkins], 'The Visitation of Dens', *Penitentiary Work in the Church of England*, pp. 126–7. See also Robinson, *The Soldier's Friend*, pp. 97–8.

[67] Robinson, *The Soldier's Friend*, pp. 98–9.

[68] Hopkins, *An English Woman's Work among Workingmen*, pp. 41–3. Ellice Hopkins, *Home Thoughts for Mothers and Mothers' Meetings*, London, 1869, p. 3.

to admit that her 'prayer seemed to have gone no higher than the ceiling'.[69] But in the following three weeks she claimed to have rescued seven girls, one of whom simply left the brothel after hearing that ladies cared enough for the likes of her to visit. Miss Hopkins, though anxious to encourage others to follow in her footsteps, did not wish to deceive women about brothel visiting. It was unpleasant work that had 'emphatically to be done on the knees'.[70] The comments of other visitors bear this out. The rude jest and the door slammed in the face were to be expected. The redoubtable Clara Lowe once found herself locked into a whore-house by some girls as a joke.[71]

Effective rescue work depended on a variety of factors, not least among them the hardness of the prostitute and the tenderness of the visitor. Undoubtedly, the volunteers had a higher rate of success among those prostitutes who were new at their trade and suffering from remorse. By her sympathetic understanding Mrs Julia Wightman, author of the celebrated *Haste to the Rescue*, a relatively subtle temperance tract, 'plucked many a brand from the burning'. She recorded the story of one of her successes, a girl named Charlotte, whom she had been asked to visit by a friend. At the brothel door she simply offered an invitation to her house. In the comfortable surroundings of her home that evening she took the girl by the hand and said: 'Don't be afraid of me. I am not going to say one unkind word,—not one word of reproach. I offer myself as a true and sincere friend to you,—as true a friend as I was to your father. You are now an orphan Charlotte. If you will leave your present course of life, I will be like a mother to you as long as you live.'[72] Charlotte, we are told, burst into tears and agreed to enter a penitentiary. Mrs Wightman spoke of 'God's mercy in Christ', clothed her, and prepared her for the disciplined regime ahead. To reduce the girl's embarrassment the penitentiary chosen was miles away, but none the less Mr

[69] [Ellice Hopkins], *The Visitation of Dens*, London, 1874, pp. 12–13.
[70] Ibid., pp. 17–18.
[71] Mary H. Steer, *Opals from Sand, a story of early days at the Bridge of Hope*, London [1912], p. 58.
[72] [Wightman], *Annals of the Rescued*, pp. 161–4.

Wightman, true to her word, kept in close touch through visits and letters. Charlotte proved a relatively easy case, for she was young and obviously unhappy with her life. Moreover, Mrs Wightman had the advantage of knowing her father through her district visiting. Her willingness to act as a substitute mother was one of the heartfelt tactics regularly employed by brothel visitors.

Tactics were much discussed. There were a few minor disagreements. Ellice Hopkins, for one, thought that the morning was the ideal visiting time. Mary Steer, who was inspired by Miss Hopkins to do rescue work in the Ratcliff Highway, preferred noon to three, for in her experience the girls were often asleep until noon; after three they were commonly drunk.[73] Some visitors thought it an advantage to go alone, for it showed a certain fearlessness; others thought it wise to go in pairs and split up once they had got inside. It was frequently recommended to go with a policeman, a health inspector, or a Bible-woman, especially in the worst districts. One school of thought believed that local working-class women made the best den visitors, for they were likely to mix more easily with the girls than ladies from distant neighbourhoods. *The Magdalen's Friend* was explicit on this issue: 'It is for the Bible women of the nineteenth century to penetrate the moral gloom of our cities and to regenerate society with the antidote to all impurity.'[74] Whatever their class, rescue workers were preferably to be over thirty, married, and innocent. Their dress, in contrast to district visitors, was to be natural and not overly plain. As one writer put it: 'The demi-semi half-sister poke bonnet severe black individual is a mistake here.'[75]

Getting into a den could be quite a problem, for brothel-keepers and landlords were predictably difficult and often obstructive. Ideally they were to be avoided. There was no point in trying to get the assistance of such people, argued the journal *Seeking and Saving*; it was best 'boldly to walk in, when the door is opened, and to go upstairs to the girls'

[73] Steer, *Opals from Sand*, p. 58. See also Mary H. Steer, 'Rescue Work by Women among Women', *Woman's Mission*, ed. Burdett-Coutts, pp. 149–59.

[74] Oct. 1860, p. 201.

[75] Brinckman, *Notes on Rescue Work*, p. 39.

rooms, without asking for permission'.[76] If a landlord comes to the door 'look him quietly and firmly in the face, . . .' advised Ellice Hopkins, and say: 'If you had a sister or daughter in one of these houses, wouldn't you like a lady to come and be a friend to her'?[77] In the case of upper-class dens with locked doors and servants, cards were commonly sent upstairs. If the door were locked and no one answered, it was not unknown for the visitors to use a little force. Nor were they above a ruse. One of Annie Macpherson's aids at the Bridge of Hope Refuge went around the brothels in Spitalfields with a petition ostensibly for the protection of shop assistants, which they pretended was to go to Parliament.[78] Such tricks were thought excusable in a cause so important.

Once inside the visitor was pretty much on her own. As with rescue work in the streets the principal object was to get the prostitute to leave her environment, and anything which had this effect was acceptable practice. As always visitors passed out tracts and cards, which listed refuges and registry offices where respectable employment might be found. But it was thought more desirable to persuade the girl to leave the brothel immediately. If at all possible a friendship was to be developed and the girl invited to the visitor's home, where she was to be drawn into the activities of the household and to be treated with 'simple human fellowship'.[79] *Seeking and Saving* highly recommended tea and piano-playing, laid on without male interference.[80] At all costs, visitors were to avoid harangues, for prostitutes normally resented being spoken down to, especially those of the higher class. If a scene did occur in a brothel, flowers and a kind note might be sent to make amends. Whatever the class of prostitute, or for that matter the class of visitor, the essential characteristics to be cultivated were 'the preparation of the heart, sound common sense, faith in God and love for souls'.[81]

[76] *Seeking and Saving*, i, May 1881, p. 8. Compare with *All the World*, x 1892, pp. 343–6.

[77] [Hopkins], *The Visitation of Dens*, pp. 20–1.

[78] Steer, *Opals from Sand*, p. 49.

[79] Ellice Hopkins, *Work in Brighton; or, Woman's Mission to Women*, London 1877, p. 98.

[80] *Seeking and Saving*, i, May 1881, p. 8.

[81] Ibid.

Common sense was not always in evidence among brothel visitors but faith in God and love for souls were rarely lacking. It is fair to say that without such attributes women rescue workers would not have been in the field. They were exceedingly anxious to point out their indebtedness to the New Testament for their inspiration. The image of Mary sitting at the feet of the Saviour holding the soiled hand of Mary Magdalene made a powerful impression on them.[82] Even more potent was the example of Jesus. 'If you would follow in the blessed steps of His most holy life', remarked a writer in *The Magdalen's Friend*, 'you must not neglect your duty to your fallen sisters'.[83] One rescue worker, who had to work herself up to her task by two years of prayer, declared after a 'cruise' in Regent Street: 'Shall we call ourselves *servants*, and refuse to obey our Master's command, when he says, "Go forth, and compel them to come in, that my house may be filled." '[84]

The imitation of Christ's compassion was their ideal. 'None other but He can know what unutterable agony goes up by day and by night from the loathsome chambers, the pestilential dens, in which these homeless, hopeless, and decaying mortals hide themselves in misery to die.'[85] Was not his tenderness toward the fallen a model of forgiveness and sanctity? Had he not cast the seven devils out of Mary Magdalene? (Luke 8: 2.) Had he not challenged the accusers of the woman taken in adultery: 'He that is without sin among you, let him first cast a stone at her.' (John 8: 7.) Had he not said: 'Neither do I condemn thee: go thy way; from henceforth sin no more.' (John 8: 11.) Such passages reinforced the view of women reformers that the unique message of Christ was the worth of the individual soul and the converting power of love. Just as Jesus hated sin but loved the sinner, they too must hate and love likewise. And while they had to call on all their faith to do so, they seldom feared the dangers or doubted the consequences. In a world in which all were sinners, in a world in which they too, but

[82] *The Magdalen's Friend*, iii, 1862, p. 174.
[83] Ibid., i, 1860, p. 202.
[84] Blackmore, *The London by Moonlight Mission*, pp. 313–16.
[85] *The Female's Friend*, March 1846, p. 61.

for the grace of God, might have fallen, there could be little moral danger in reaching out to these poor, despised women. The stakes were high, for as St. Paul warned: 'If ye live after the flesh, ye shall die; but if by the spirit ye mortify the deeds of the body, ye shall live.' (Romans 8: 13.)

The message that through Jesus there was escape into everlasting life could be seductive. However hardened, few prostitutes were altogether free from remorse. They would have to have been very self-possessed indeed to be wholly unaffected by the prevailing moral tone of the world outside their brothels. Most of them, it should be said, did not choose their profession as one chooses to be a hairdresser or a historian. As Mayhew pointed out, they were usually reduced to it by their social and economic circumstances. And as the rescue workers discovered, many prostitutes were in misery, ridden with guilt, and anxious to reform. The fact that thousands of them voluntarily entered homes and refuges with stringent regulations bears this out. As we have seen, the inculcation of a sense of guilt was a dominant part of the indoctrination in such institutions. In so far as prostitutes felt a sense of shame, the life of Christ offered them an escape, however arduous. Through his tender mercies they might find salvation in the hereafter, and through his benevolent agents they might find a respectable trade in the here and now.

Rebecca Jarrett, seduced at the age of twelve, was one of the thousands of fallen women who found her feet under the consoling influence of Christ and his agents. Her autobiography opens: 'This is written by my own self, not to boast of my disgusting life, *no*, but to show how good *Jesus* is to a poor, lost, degraded woman. I do earnestly ask God's blessing on each one who takes the poor, fallen one by the hand and helps to raise them up.'[86] She had been raised up by the hands of Catherine Booth and Josephine Butler, under whose influence she worked as a brothel visitor herself in Winchester and Portsmouth. She was probably the most widely known 'fallen woman' turned rescue worker in the nineteenth century. At the end of her prison sentence, which

[86] Quoted in Charles Terrot, *The Maiden Tribute*, London, 1959, p. 96.

stemmed from the celebrated abduction of Eliza Armstrong, she was met by Mrs Bramwell Booth and, under a different name, lived as a Salvationist for forty years.[87]

No charitable activity brought women of different backgrounds into more intimate contact than rescue work. This is not surprising for the volunteers commonly argued that the best hope for their fallen sisters lay in their friendship with respectable women.[88] The case of Rebecca Jarrett and others suggest that this was not simply fatuous talk, though, to be sure, relations were often strained between rescuer and rescued, particularly once the latter found herself in a public or charitable institution. The well-born Josephine Butler was the constant companion of prostitutes. 'Possessed with an irresistible desire to go forth and find some pain keener than [her] own' after the sudden death of her daughter, she roamed Liverpool in search of ailing prostitutes.[89] With the support of her understanding husband, she brought them home to die. Her first patient, Marion, found in a Magdalene ward, lived for three months with the Butlers, mastering the New Testament, before she died of a lung disease. As Mrs Butler put it, Marion had been ruined in a gentleman's house, she could at least die in the home of another gentleman.[90] 'Numberless' other diseased and dying women followed Marion into the house over the years, as many as five at any one time.[91] Through her friendship with these pitiable women, Mrs Butler assuaged her own suffering and attempted to live a life in accord with the social doctrine of Christ, which she interpreted as the principle of perfect social equality.

To Mrs Butler and many other female rescue workers the principle of social equality had special reference to the status of women. 'Search throughout the Gospel history', she

[87] Coutts, *No Discharge in this War*, p. 112. See also Josephine Butler, *Rebecca Jarrett*, London [1885].

[88] See, for example, Hopkins, *Work in Brighton*, p. 98.

[89] *Josephine E. Butler, an Autobiographical Memoir*, eds. George W. and Lucy A. Johnson, Bristol and London, 1909, pp. 58–9.

[90] Ibid., p. 67.

[91] *P. P., Reports from Commissioners, Report of the Royal Commission upon the Administration and Operation of the Contagious Diseases Acts*, 1871, xix, p. 437.

once wrote, 'and observe [Christ's] conduct in regard to women, and it will be found that the word liberation expresses, above all others, the act which changed the whole life and character and position of the women dealt with, and which ought to have changed the character of men's treatment of women from that time forward'.[92] Few things disturbed women like Mrs Butler more than the glaring inequalities between the sexes which fornication and prostitution highlighted. The double standard of morality, by which society punished women for deeds that it excused in men, appalled them. The misery and vice of prostitution was a reflection of society's disrespect for women and family life; and given the conventional view of woman's nature, it was a reflection of society's hypocrisy. Women reformers saw their sex as distinct from men to be sure, but not less worthy of equal treatment; and they would not rest until they got it. On the issue of social purity, wrote Ellice Hopkins, 'we have let men impose their views on us too long. . . . We have too long forgotten, that if to men have been given the bodily strength and the intellectual pre-eminence, it is the woman who is the conscience of the world'.[93]

Nor did they doubt what needed to be done: 'We can unite in our common redeemer in defence of our common womanhood', declared Ellice Hopkins.[94] Thus women carried the banner of moral reform and social purity across the country. They wrote and distributed pamphlets, stirred up husbands, MPs, ministers, journalists, magistrates, and the police, petitioned Parliament, spoke to conferences and gatherings of working men, added their weight to the charitable societies in sympathy with their aims, corresponded with each other, and set up countless women's associations. 'Let us remember we are going up against a great organised evil', warned Miss Hopkins, 'and we can only meet it by counter-organisation'.[95] It was estimated that well over two hundred rescue and moral reform charities resulted from

[92] Butler, ed. *Woman's Work and Woman's Culture*, p. lix; quoted in *Josephine E. Butler*, eds. Johnson and Johnson, p. 85.

[93] Hopkins, *Work in Brighton*, pp. 40–1.

[94] Ellice Hopkins, *How to Start Preventive Work*, London, 1884, p. 37.

[95] Hopkins, *Grave Moral Questions*, p. 28.

her driving spirit and organizational flair alone.[96] 'Our principle', she announced, 'is to work at the centre of this great evil, not merely at the extremity, to attack causes not only to cure results'.[97]

Such a principle pushed many women rescue workers, agitated by the constant reminders of their sex's disadvantages, in directions which they could not easily have anticipated. The campaign against the Contagious Diseases Acts was one of the movements highlighted and greatly influenced by rescue work. Indeed, the involvement of women in this battle grew directly out of their work with prostitutes. Passed in the 1860s, the Acts sought to reduce venereal disease in the armed forces. They empowered the police to pick up, register, and have medically examined women suspected of prostitution in port and garrison towns. Those women found to have contracted venereal disease could be detained in a certified Lock hospital for up to nine months.[98] Those found free of disease none the less remained on the police register and were required to return for periodical examinations. Refusal to submit to the medical examination, or to treatment in the case of those declared diseased, was punishable by imprisonment for no more than one month in the case of first offenders and up to three months for subsequent offences. One effect of the Acts was the creation of an 'outcast' and more professional class of prostitute, 'Queen's women' as they were sometimes called.[99] Another

[96] Rosa M. Barrett, *Ellice Hopkins: a memoir*, London, 1907, p. 5.

[97] Ellice Hopkins, *Ladies' Associations for the Care of Friendless Girls*, London, 1878, pp. 19–20.

[98] The Act of 1869 extended the period of detention, which had been set at six months by the Act of 1866. The Acts are reproduced in *The Shield*, April 4, 1870, pp. 36–7, 40–1.

[99] *P. P., Reports from Commissioners, Report of the Royal Commission upon the Administration and Operation of the Contagious Diseases Acts*, 1871, xix, p. 429; *Reports from Committees, Report from the Select Committee on Contagious Diseases Acts*, 1882, ix, p. 239. There is an interesting discussion of this issue in Judith Walkowitz, 'The Making of an Outcast Group: Prostitutes and Working Women in Nineteenth-Century Plymouth and Southampton', *A Widening Sphere*, ed. Martha Vicinus, Bloomington and London, 1977, pp. 72–93. See also Judith R. Walkowitz and Daniel J. Walkowitz, ' "We are not Beasts of the Field": Prostitution and the Poor in Plymouth and Southampton under the Contagious Diseases Acts', *Clio's Consciousness Raised*, eds. Mary S. Hartman and Lois Banner, New York, 1974, pp. 192–225.

was the creation of charities determined to end this state of affairs.

Reformers of both sexes led a spirited opposition to the Acts, which they believed to represent a sinister double standard of morality and an infringement of the individual liberty of women. Many of the working men who joined the campaign thought the Acts to be an embodiment of class legislation, that is they saw them as 'disinfecting' working-class girls for the use of the army and the upper classes. For women the decision to do battle with the government was not taken lightly. Mrs Butler agonized over it: 'If doubt were gone, and I felt sure He means me to rise in revolt and rebellion (for that it must be) against men, even against our rulers, then I would do it with zeal, however repulsive to others may seem the task.'[100] Coming to the conclusion that Jesus was sympathetic to female rebellion, she led the assault as the Secretary of the Ladies' National Association for the Repeal of the Contagious Diseases Acts, established in 1869. By 1882 it boasted ninety-two local committees, making it one of the larger charities run by women in the second half of the century.[101] It was also an excellent example of what Ellice Hopkins, who supported the cause, had in mind when she called on women to organize in defence of their sex. The name chosen for the society's periodical, *The Shield*, was telling; it came from Paul to the Ephesians: 'The Shield of faith, wherewith ye shall be able to quench all the fiery darts of the wicked.'

Not all women rescue workers joined the crusade against the Acts however. The reason becomes clear from one feature of their administration. For in carrying out their duties under the law the police and health inspectors often worked hand in hand with rescue workers, sometimes to good effect. As the Select Committee of the House of Lords reported in 1882: the Acts 'make an attempt at reclamation by moral and religious agencies an essential part of every attempt to

[100] From her manuscript book, September 1869, quoted in *Josephine E. Butler*, eds. Johnson and Johnson, p. 91.

[101] *P. P., Reports from Committees, Report from the Select Committee on Contagious Diseases Acts*, 1882, ix, p. 230.

check the evils of prostitution'.[102] The police were well known to brothel visitors, who occasionally made their rounds with them, and to the superintendents of refuges and Lock hospitals, to whom they frequently referred prostitutes. In Aldershot, for example, the police were in touch with Miss Daniell's Home for Girls.[103] Consequently the managers of such institutions and the corps of street and brothel visitors were far from unanimous in condemning the Acts. Miss Mary Anne Webb, the superintendent of the Lock Hospital in Chatham and formerly a deaconess, testified before the Select Committee in 1882 that they extended her power to do good.[104] Mrs Colebrook, a rescue worker in Portsmouth and manager of the Southsea Home for Girls, called the Acts 'a very useful auxiliary' before a Royal Commission some years before.[105] Such testimony was very disturbing, not to say damaging, to Mrs Butler's cause.

Women rescue workers were then not of one mind about the Contagious Diseases Acts. But for many of them, probably the majority if the testimony in the *Parliamentary Papers* is anything to go by, the disadvantages of the Acts outweighed any benefits of a referral service provided by the police. Nor did the widespread support for the Acts from the medical profession do much to change the views of those to whom religious and libertarian considerations were paramount.[106] When the physician Elizabeth Garrett (later Anderson) called for an extension of the Acts in 1870, she created a stir among the officers of the Ladies' National Association, but her remarks did nothing to deter

[102] *P. P., Reports from Committees, Report from the Select Committee on Contagious Diseases Acts*, 1882, ix, p. xx.

[103] Ibid., *Accounts and Papers, Copy of Annual Report, for 1875, of Captain Harris, Assistant Commissioner of Police of the Metropolis, on the Operation of the Contagious Diseases Acts*, 1876, lxi, p. 8. The name is spelled Daniel in this report.

[104] Ibid., *Reports from Committees, Report from the Select Committee on Contagious Diseases Acts*, 1882, ix, p. 414.

[105] Ibid., *Reports from Commissioners, Report of the Royal Commission upon the Administration and Operation of the Contagious Diseases Acts*, 1871, xix, p. 409.

[106] In addition to the evidence from doctors in the *Parliamentary Papers* on the Contagious Diseases Acts, their views are expressed in some detail in the Public Record Office, HO 45 9511/17273A.

their campaign.[107] They were, or so they believed, battling the anti-Christ, and consequently no argument in favour of the Acts could be of any force. Where doctors saw disease they tended to see sin; and in their minds the Acts, in making sin more convenient, encouraged lasciviousness, or death.

At the heart of their opposition was a biblical literalism, common to many Victorians, which was awesome in its implications. The Wesleyan Conference sent a petition to the House of Commons describing the Acts as 'tending to facilitate the habitual breach of the holy laws of Almighty God'.[108] The remark of a brothel visitor from Birmingham, Mrs Lewis, to the Royal Commission in 1871, was more specific: 'My Bible tells me I am not to fulfil the lusts of the flesh, and that the wages of sin is death, and therefore to provide for prostitution getting the better of God's law of death was to me a terrible sin, and such a law to my mind as a Christian woman and a mother was a bad law.'[109] Mrs Butler, while not altogether insensitive to the medical side of the issue, placed great store by the healing powers of prayer. In a letter to James Wilkinson, Swedenborgian and homoeopathic doctor, she claimed that when her girls feared the ordeal of the government medical examination for venereal disease 'I have asked them to pray, and we all prayed, and they got well without any examination.'[110]

Mrs Butler was rather better at marshaling the Christian sentiment of women than in persuading Parliament to abolish the despised Acts. When they were suspended in 1883 and finally repealed in 1886, much of the credit was due to her colleague James Stansfeld, who as the head of various government departments knew the ins and outs of politics. He set up an effective parliamentary inquiry into the Acts and in doing so put the argument for repeal 'on its legs'.[111] None

[107] *The Shield*, March 7, 1870, p. 2.

[108] *The Shield*, Oct. 1, 1870, p. 242.

[109] *P. P., Reports from Commissioners, Report of the Royal Commission upon the Administration and Operation of the Contagious Diseases Acts*, 1871, xix, p. 426.

[110] James John Garth Wilkinson, *The Forcible Introspection of Women for the Army and Navy by the Oligarchy, considered Physically*, London, 1870, p. 25.

[111] J. L. and Barbara Hammond, *James Stansfeld. A Victorian Champion of Sex Equality*, London, 1932, pp. 293–4.

the less Mrs Butler's crusade made an impact, particularly in the early years; and the repercussions of her movement went far beyond the immediate repeal campaign. The Acts were an illustration of the insensitivity of men, most notably the nation's governors, to women's feelings; and the resulting outrage stimulated the movement for legal reform and for social purity, for women doctors and for women's suffrage. The connections, both of personnel and purpose, between the Ladies' National Association and the Vigilance Association for the Defence of Personal Rights, the Social Purity Alliance, the Moral Reform Union, the National Vigilance Association, the National Society for Women's Suffrage, the Ladies' Association for the Care of Friendless Girls, and other charities in which women played a prominent part, were intimate.[112]

Rescue work stimulated all of these various campaigns, and none of them more than the movement for the legal protection of women and children. While rescue workers were divided over the repeal of the Contagious Diseases Acts, they were unanimous in supporting legislation which would check the cause of prostitution and punish the seducers of and trafficers in young girls. They pointed an accusing finger at the legal system which allowed brothels to operate and which placed women and children at the mercy of procurers, perverts, and old roués. The law, for example, did not give a parent the power to enter a brothel and remove a daughter, though 'if the woman who kept the establishment had stolen his spoons instead of his daughter he could have gone to a magistrate and taken out a search-warrant, and entered the house and recovered the stolen property'.[113]

In the mid-nineteenth century no law existed to punish a procurer, and though brothel-keeping was indictable at common law, the legal proceedings necessary to gain conviction were expensive and time-consuming. The London Society for the Protection of Young Females was one charity

[112] Most of these connections have been traced in Brian Harrison, 'State Intervention and Moral Reform in nineteenth-century England', *Pressure from Without in early Victorian England*, ed. Patricia Hollis, New York, 1974, p. 319.

[113] Quote from the Bishop of Peterborough in Ellice Hopkins, *The White Cross Army*, London, [1883], p. 1.

which had had some success in indicting brother-keepers from the mid-1830s, but it was well aware of the limitations of the law.[114] So too was the St. George's Vigilance Society, Hanover Square, which once applied to the Home Office, unsuccessfully it seems, for the fines imposed on brothel-keepers to be appropriated for its own use.[115] Nor were the police happy. Sir Richard Mayne, Commissioner of the Metropolitan Police, admitted that they could do little more than observe brothels, collect information, and give evidence in support of indictments brought by parish authorities and rescue workers.[116]

Women reformers showed a particular concern for the legal condition of the many young girls exposed to the conditions of prostitution: children exploited by their parents to bring in extra earnings; servants defiled by their masters and reduced to the streets; and the children of prostitutes and brothel-keepers themselves. 'I have myself had in my hands a child of seven', reported Ellice Hopkins,

> who was sold by her own mother, in the public den of infamy kept by her; who had passed through a public Lock hospital, and whom I found in a public penitentiary, sitting up to a table among a number of abandoned women—as sweet a blossom of a child as you would wish to see—with her little thimble and needle and thread hemming a duster, the first use of which should have been to strangle the men who had degraded her.[117]

In this case one of the defilers was caught, only to be released by the judge, who argued that as the man was not the first to degrade the child, he could not be punished under the law. 'Is it the will of the British public that little

[114] *Fourth Report of the London Society for the Protection of Young Females*, London, 1839, pp. 10–13. By 1860 this society claimed to have suppressed over 500 brothels. *Magdalen's Friend* Oct. 1860, i, p. 216. The Associate Institute for Improving and Enforcing the Laws for the Protection of Women also had some success in carrying out prosecutions of brothel-keepers. See *Meliora*, i, 1858, p. 72.

[115] Public Record Office HO 45 9963/X14891.

[116] Ibid., HO 45 6628/3 and 7370/11. Another tactic was for the police to call in the Inland Revenue, to prosecute unlicensed refreshment houses or 'night houses'. This was not always successful, for such establishments were on the alert for such visitors and barred their entry. See HO 45 9511/17216/14–15.

[117] Ellice Hopkins, *Drawn unto Death*, London, 1884, p. 9.

children of seven should be sold in a den of infamy', asked Miss Hopkins?

> Is it the will of the men and women of Christian England that little girls of twelve should have their tender childish bodies beaten black and blue because they refused to be the victims of devils—men I won't call them? Is it the will of the nation that the national funds should not be spent in saving her girl-children from being brought up for degradation and becoming fountains of corruption . . . ? Is it the will of the community that this monstrous anomaly should go on; houses declared illegal by the English law, but in which our wretched English *laissez faire* allows young life to be brought up?[118]

Miss Hopkins, for one, battled relentlessly to end this anomaly. In what is an excellent example of the ability of women to effect government policy, she was successful in getting the Industrial Schools Act of 1866 amended. By this Act a person could bring before a magistrate any child under fourteen who was found begging, destitute, homeless, or in the company of thieves. The magistrate had the authority to send the child to a certified industrial school if he thought any of these descriptions were satisfied. Miss Hopkins wanted any child found in the company of prostitutes or residing in a brothel to be given the same treatment. In 1880 she got her wish, in what was to be hailed as the Ellice Hopkins' Act by her followers, an encomium not undeserved.[119] But as the Select Committee of the House of Lords on the Law relating to the Protection of Young Girls discovered in 1882, she was still unhappy. In her testimony she accused the magistracy of being lax in administering the Act; and she made several proposals, which included extending legal protection to girls up to twenty-one.[120]

Indignant females, aroused in large part by rescue workers,

[118] Ibid., pp. 11–12.

[119] She testified that she had 'originated' the Act of 1880. *P. P., Report from Committees, Report from the Select Committee of the House of Lords on the Law relating to the Protection of Young Girls*, 1882, xiii, p. 6. The Acts, 29 and 30 Vict. c. 118 and 43 and 44 Vict. c. 15 are cited in Hopkins, *Drawn unto Death*, pp. 23–4. See also, *Fifty Years' Record of Child-Saving and Reformatory Work*, p. 21.

[120] *P. P., Reports from Committees, Report from the Select Committee of the House of Lords on the Law relating to the Protection of Young Girls*, 1882, xiii, pp. 9–11. She conceded that eighteen would probably give adequate protection.

were never more determined to protect women and children from the dangers of prostitution than in the early 1880s. Ellice Hopkins was in the thick of it. She collected information, wrote letters and pamphlets, testified before the House of Lords Select Committee, and worked with and established charitable associations dedicated to amending defective legislation. In 1882 she called on women from around the country to deluge Parliament with petitions, which would put pressure on the House of Commons to give effect to the proposals coming out of the Select Committee Report of the House of Lords.[121] She must have been pleased when Lord Shaftesbury, a member of the Committee, asked another witness, Mrs Wilkes of Poplar, if she had seen the vast number of petitions from women 'praying for equal laws, as between the sexes'?[122] Thousands of female signatures poured into Parliament: they came from parishes fed up with street-walkers and 'refreshment houses'; from societies like the Ladies' Association for the Care of Friendless Girls, the YWCA, and the Metropolitan Association for Befriending Young Servants; and from individuals like Baroness Burdett-Coutts and Emily Janes, editor of *The Year Book of Woman's Work*, who wrote on behalf of other charities committed to legal reform.[123] It was a remarkable display of female sentiment and it kept the subject of the law regarding women and children in the minds of MPs and Home Office officials.

Despite their campaign, the Criminal Law Amendment Bill, supported by Lord Shaftesbury, ran into difficulties in the House of Commons. By the end of May 1885, it appeared to be a dead issue. Uneasy about the Bill's prospects, the reforming journalist William Stead, assisted by Mrs Butler and rescue workers of the Salvation Army, tried a new tactic. Stead had previously given over his columns in the *Pall Mall Gazette* to Mrs Butler. Now he took up the pen himself and created his own headlines by having a child, Eliza

[121] Hopkins, *Grave Moral Questions*, pp. 10–13.

[122] *P. P., Reports from Committees, Report from the Select Committee of the House of Lords on the Law relating to the Protection of Young Girls*, 1882, xiii, p. 41.

[123] See, for example, Public Record Office, HO 45 9546/59343G *passim.*

Armstrong, bought for immoral purposes by the former procuress Rebecca Jarrett. This story has been told at length elsewhere;[124] it is only necessary to say that in exposing the horrifying details of child prostitution and sexual abuse Stead stirred up an already sensitive public. The pressure on Parliament mounted quickly. For its part the Salvation Army, inspired by a speech of Catherine Booth, collected 393,000 signatures in seventeen days; eight cadets carried this weighty petition, which demanded legal protection for the nation's children, into the House of Commons.[125] With its interest in the Criminal Law Amendment Bill rekindled, Parliament passed it on August 14, 1885. Among its clauses the Act gave further protection to women and girls from procurers, raised the age of consent to sixteen, and provided for the summary conviction of brothel-keepers.[126]

But this was not the end of the matter for women rescue workers and moral reformers. Certainly most of them were pleased with the new Act. In Manchester, for example, Mrs Arthur Weigall and her aids, working with the police and magistrates, found it easier to close down brothels and clear the streets of prostitutes than previously.[127] Yet to many women, Ellice Hopkins and Josephine Butler included, the law was far from perfect on the protection it afforded women and children. They joined the National Vigilance Association, formed in 1885 as a result of Stead's revelations, to monitor the Act's adminstration. In the Preventive Sub-Committee Miss Hopkins was joined by, among others, Mrs Henry Fawcett, Mrs Bramwell Booth, Mrs Ormiston Chant, and Mary Steer. They collected information and had short abstracts of the Act distributed to the police, usually through ladies who attended police classes. The Legal Sub-Committee, with Miss Hopkins again in evidence, undertook prosecutions under the new Act and promoted further amendments of the law. It was particularly anxious that the courts convict the 'wealthy scoundrels who systematically commit crimes against young girls'.[128]

[124] See, for example, Terrot, *Maiden Tribute.*

[125] Coutts, *No Discharge in this War*, pp. 109–10.

[126] 48 and 49 Vict. c. 69.

[127] Weigall, *Seeking and Saving*, pp. 22–3.

[128] *National Vigilance Association. Report of the Executive Committee*, London, 1887, p. 2, *passim.*

In its intensity the moral reform campaign of the later nineteenth century, in which the National Vigilance Association was prominent, recalls the battle with immorality which took place in the days of William Wilberforce. The moral reform charities founded in the 1870s and 1880s, however, appear rather less class-biased than their late eighteenth- and early nineteenth-century predecessors, perhaps because they operated in more settled times. They were, of course, in the tradition of the Society for the Suppression of Vice, whose attacks on the pastimes of the poor have been frequently noted. Indeed, the National Vigilance Association received funds from the Vice Society, which was still active in the 1880s.[129] Yet the late nineteenth-century moral reform movement also owed a great debt to the tradition of Hannah More, whose attacks on the manners and morals of the rich have gone relatively unnoticed. As we have seen, rescue and social purity workers from mid-century onwards were as anxious about the souls of the high-class prostitutes as the low-class ones; nor did they wish to discriminate between the rich and poor 'scoundrels', who kept such women in business.

The societies established by rescue and social purity workers in the late nineteenth century are not easy to label. They came in all shapes and sizes and varied considerably in their methods. The Association of Men for the Defence of Women from Dishonour, for example, did not greatly resemble the Woman's League, which drew together a host of prominent ladies dedicated to bringing all classes of society together on issues touching marriage and family life.[130] Nor did the White Cross Army draw on the same social groups or use the same methods as the Society for the Suppression of the Traffic in Girls. Yet most of these charities, whatever their size or make-up, had at least one object in common. It was summed up by the principle of the Vigilance Association for the Defence of Personal Rights, founded in 1871: to promote the equality of all persons before the law, regardless of sex or class.[131]

[129] Ibid., p. 26.
[130] *Seeking and Saving*, ii, Feb. 1882, p. 28; v, Oct. 1889, pp. 439-43.
[131] *Twelfth Annual Report of the Vigilance Association*, London [1883], p. 2.

In one of the more intriguing moral reform projects of the century Ellice Hopkins carried the battle into the camp of men. This was the White Cross Army, which she established in 1883 with the help of the Bishop of Durham. Her association with the Church of England was intimate, and in the 1870s she made a plea for it to take a wider interest in moral reform, with special reference to the prevention of prostitution.[132] The Church could not claim to be a leader in this field, despite its various penitentiaries and the rescue work of the deaconesses. The White Cross Army represented a fresh offensive. It recruited only men, who, as Miss Hopkins saw it, were to be in the front line of social purity. These chastened foot-soldiers pledged to show a 'chivalrous respect for womanhood', to extend the law of purity equally to men and women, and to forgo foul language and indecent behaviour.[133] The society they joined promised to support morality sermons, midnight meetings, and sought legal reform for the protection of women and children.[134]

As with most moral reform societies it is difficult to judge the success of the White Cross Army, or White Cross League as it came to be called in 1891, when it amalgamated with the Church of England Purity Society. We do know that a large number of artisans and miners in Yorkshire signed up and that there were branches in Army barracks and ships of the Royal Navy, and in all but two dioceses. In time it spread to most parts of the British Empire and to other areas of the world which felt in need of its cleansing influence— a branch opened in Chicago in 1885. Miss Hopkins, who read Bunyan and the sermons of C. H. Spurgeon to improve her public speaking, addressed meetings of up to 2,000 working men at a time; and she wrote most of the pamphlets which made up the 'White Cross Series', which sold over two million copies in England and America by the end of the

[132] Ellice Hopkins, *A Plea for the Wider Action of the Church of England in the Prevention of the Degradation of Women*, London, 1879.

[133] There is an example of a White Cross Army pledge and membership card in the British Library under the reference White Cross Army.

[134] *White Cross League Church of England Society Twelfth Annual Report*, London, 1895; Ellice Hopkins, *The White Cross Army*, London [1883]; Ellice Hopkins, *The White Cross Army. A statement of the Bishop of Durham's Movement*, London, 1883.

century.[135] One of her effusions, 'True Manliness', sold 300,000 copies in its year of publication.[136] All in all, the Army was a peculiarly Victorian attempt to get men to mend their ways and may be seen as part of the moral arsenal of the Empire. For our purposes it shows the breadth of female charitable activity and how women sought to strengthen the view that they were both special and worthy of equal treatment.

The moral reform societies more familiar to us were very wide-ranging in their interests. In addition to monitoring the Criminal Law Amendment Act, the National Vigilance Association promoted workhouse visiting, rescue work, and the election of women as Poor-Law Guardians; it objected to theatres, to music halls, to massage houses, and to Charles Dilke; and in one scheme which bore the trade mark of Miss Hopkins, it made an effort 'to get men to visit the fallen of their own sex in Lock hospitals'.[137] The Social Purity Alliance, established in 1873, sponsored improving lectures and advocated temperance, decent dress, legal reform, and 'one standard of morality for men and women'.[138] For its part, the Moral Reform Union, formed in 1882, supported emigration, various legislative amendments, and the election of female Poor-Law Guardians; and it worried about the ballet, the state of public parks, and 'the conduct of boys and girls who meet in railway carriages on their way to school'.[139] Like the other societies of its type it was 'anxious to keep true to the pole of principle—equal justice for men and for women'.[140]

Such preoccupations showed the unmistakable signs of women's influence. When they did not manage the moral reform charities outright, which was common enough, they

[135] Ellice Hopkins, *The Power of Womanhood*, London, 1899, back cover; *The Times*, 24 August, 1904.

[136] Ellice Hopkins, *Dictionary of National Biography*.

[137] *National Vigilance Association. Report of the Executive Committee*, 1887, pp. 2, 14–15, 25. See also William Alexander Coote, *A Romance of Philanthropy*, London, 1916.

[138] *Social Purity Alliance. Established in 1873. Annual Report*, London, 1880, pp. 20–1.

[139] *The Moral Reform Union. Second Annual Report*, pp. 14–15; *Fifth Annual Report*, p. 5.

[140] Ibid., *Fourth Annual Report*, p. 10.

were a powerful force in them. The Church of England Purity Society was exceptional in excluding women from its council (a 'fatal mistake' remarked a spokesman for the Moral Reform Union).[141] In the Moral Reform Union, women, who made up three-fourths of the subscribers, not only joined the Committee; they audited the books and did secretarial work. In the National Vigilance Association, where they contributed about half of the funds, they were well represented on the Council and dominated most of the sub-committees. They were also indispensable in the Social Purity Alliance, not least in giving direction and a sense of urgency to the operations. Would William T. Malleson, Chairman of the SPA, have come to this conclusion without a push from the ladies, perhaps from his wife, who worked by his side on the Committee?

> It is the rescue and reform of men, that is really the question of our society today. . . . And for the rescue and reform of men we look to the power and devotion of women. . . . It is to women as mothers, to women as sisters, to women in private life, and in public life, that we must look for the new force on this question, to give a new impulse, to take the matter into their own hands in order to save Society, to save England, to save India, and I almost say to save the future of the world.[142]

The moral reform societies of the late nineteenth century wanted nothing less than the regeneration of English family life through the transformation of the relations between the sexes. The Moral Reform Union stated simply that it 'was established in the interests of pure family life'.[143] If their predecessors in the early years of the century had focused their attention on the manners and morals of the working classes, they focused their attention on the manners and morals of men, whatever their class. They were a natural extension of female rescue work and an example of the way in which women were changing the character and direction of philanthropy.

James Hinton, the friend and advisor of Ellice Hopkins, once

[141] Ibid., p. 13.
[142] *Social Purity Alliance, Established 1873. Annual Report*, 1880, p. 22.
[143] *The Moral Reform Union. Fourth Annual Report*, p. 3.

told her: 'You women have been living in a dreamland of your own; but dare to live in this poor disordered world of God's, and it will work out in you a better goodness than your own.'[144] By the end of the nineteenth century few people had a finer appreciation of this 'poor disordered world of God's' than women rescue workers and moral reformers. They concluded that the most profound problems of their day had much to do with the corrupt and immoral relations between the sexes. 'The core of all social reformation', remarked Anna Jameson in her influential essay *The Communion of Labour*, '. . . lies in the working of the man and the woman together, in mutual trust, love, and reverence'.[145] Miss Hopkins and Mrs Fawcett seconded her opinion;[146] and vigilant women across the country, whether they worked as brothel visitors in provincial towns or on the committees of great moral reform societies in London, nodded their assent.

The purification of relations between the sexes was no small challenge, and many female rescue workers and moral reformers, just how many is uncertain, concluded that votes for women would further their cause. Josephine Butler, Catherine Booth, Mrs Bramwell Booth, Frances Power Cobbe, Ellice Hopkins, Margaret Lucas, Mary Steer, and others like them took up the suffrage, often against their conservative instincts, in the hope that it would lead to an extension of women's power to transform England into a more humane society. Mrs Butler typified these reformers in her belief that the vote would permit women to extend their domestic influence and force legislators to take into account the moral feelings of women.[147] Nor were such women slow to notice the irony in their situation. As the temperance worker Mary Anne Clarke put it, 'We see a well-educated woman with wealth and property at her command classed by the law with minors, idiots, and felons, while the man

[144] Hopkins, *The Power of Womanhood*, p. 9.

[145] Anna Jameson, *Sisters of Charity and the Communion of Labour*, London, 1859, p. 148.

[146] Hopkins, *The Power of Womanhood*, p. 163.

[147] For the views of the Ladies' National Association on female suffrage see *The Shield*, May 2, 1870, p. 74. Butler, ed. *Woman's Work and Woman's Culture*, pp. xvii–xviii.

who opens her carriage door or drives her horses may have a voice in the legislation of the country, be he ignorant, drunken, or depraved.'[148]

When she added her voice to the female suffrage movement, Ellice Hopkins wanted, above all, women to have the power of legislating for their own protection.[149] She justified the claims of women to vote with a very revealing argument. She compared the state with the family and suggested that as women had evolved a more prominent place within the home, 'realising the idea of Christianity', they should be more prominent by the side of men in national affairs. 'May we not find in the larger family of the State, that the work of the world is best done by the man and the woman together, each supplying what is lacking to the other, the man the head of the woman, the woman the heart of the man?'[150] For her, as for others in her camp, votes for women were simply a means by which society might be made more compassionate, by which the relations of men and women might be made equal and holy. In a world in which women had political power men might be made to rise to the standard which women set for them.

Although women rescue workers and moral reformers were among those who took up the banner of political reform, few of them were critical of the class structure and economic order of Victorian society. They were largely insensitive to the issue of whether or not the material relations and commercial values of capitalism were the root cause of the corruption in family life and the inequalities between the sexes. One wonders what they made of the treatment of impure women in the novels of Dickens. They were not unaware that prostitution, like philanthropy, flourishes in societies with inequalities of wealth. But believing that prostitution was essentially a moral problem they did not espouse a radical upheaval of the economic order. Socialist doctrines had little charm for charitable women, whose work was a

[148] *Opinions of Women on Women's Suffrage*, London, 1879, p. 56.

[149] Hopkins, *The Power of Womanhood*, p. 184. See also Frances Power Cobbe, *Why Women desire the Franchise*, London [1877], p. 3; Frances Power Cobbe, *Life of Frances Power Cobbe*, 2 vols., London, 1894, ii, p. 209.

[150] *Opinions of Women on Women's Suffrage*, p. 53.

confirmation of the existing system of class relations and whose training and inclinations were in the moral sphere. But then socialist remedies held little charm for most of the British, especially the economists.

Few women rescue workers in the 1850s could have foreseen that their concern for prostitutes, indeed their anger over their plight, would lead them to petitioning Parliament in the cause of legal reform, or lead some of them to votes for women. And as with district and institutional visiting, it is rather difficult to tell who benefited most from their labours, the objects of their sympathy, or themselves. Undoubtedly, rescue work and its corollary of moral reform gave a great boost to the organizational talents of women and heightened their sense of moral purpose. No less important, it reinforced their belief that they had a special mission in society. By the end of the century women workers in the cause had a vast and sophisticated network of connections; and because their work brought them face to face with the disadvantages of their sex, it bound them closer together than ever before.

From her experience in the anti-Contagious Diseases Acts crusade Mrs Butler concluded: 'Women are called to be a great power in the future, and by this terrible blow which fell upon us, forcing us to leave our privacy and bind ourselves together for our less fortunate sisters, we have passed through an education—a noble education. God has prepared in us, in the women of the world, a force for all future causes which are great and just.'[151] Her colleague Ellice Hopkins agreed. She had come a long way from those early days in Cambridgeshire among navvies and rude clerics and from that sense of isolation which made her once lament that 'it was hard that the power which would have been a glory to me if I were a man, would be held a shame and a disgrace to me because I was a woman'.[152] Towards the end of her life she looked back on the work which had made her make so many sacrifices and took particular pride in the growth of 'a sense of a common womanhood, that *esprit de corps* in which

[151] Quote from 'The Bright Side of the Question', in *Josephine E. Butler*, eds. Johnson and Johnson, p. 181.

[152] Hopkins, *An English Woman's Work among Workingmen*, p. 43.

hitherto we have been so grievously lacking, and which is now beginning to bind all our efforts into one great whole'.[153] And like Mrs Butler she did not doubt the source of this 'common womanhood'. Remembering the cross that biblical women had to bear, she declared:

> And now, when at the end of the ages He once again calls us women to stand heart to heart with Him in a great redemptive purpose, shall we hang back? Shall we not rather obey the Divine call, enduring the Cross, despising the shame, and, like those women of old, winning for ourselves, by faithfulness unto death, the joy of being made the messengers of a higher and risen life to the world.[154]

[153] Hopkins, *The Power of Womanhood*, pp. 2–3.
[154] Ibid., p. 190.

Conclusion

Women in late Victorian society had very many more opportunities and considerably more freedom of action than their grandmothers had had. This, of course, reflected the public's growing awareness, through such means as the popularization of the mid-century censuses, that the conditions of life had changed for women. But if we are to isolate one profession that did more than any other to enlarge the horizon of women in nineteenth-century England, it would have to be the profession of charity. It was more often unpaid than paid, but to the many women for whom financial rewards were of secondary importance it was a profession none the less. Their object was to serve, to be useful; to them philanthropy was a reflection of virtue and, more pragmatically, an escape from boredom. In an age in which women found so many doors closed, they discovered a crack in the doors of the charitable societies. With a quickness of perception for which they were supposed to be noted, they spotted their opportunity and made the most of it.

By the end of the century the female charitable reformers had broken down much, though far from all, of the prejudice that impeded their energies. They did this through persistence, tact, the persuasiveness of their financial contributions, and a willingness to do work which men found disagreeable or for which they were ill-equipped. 'Masculine officialism' was a formidable obstacle to the progress of women in organized philanthropy; even at home, of course, it took great strength of character for a wife or daughter to make headway against the wishes of her husband or her parents. It would be a distortion, however, to suppose that female philanthropists succeeded without any support whatsoever from men. Octavia Hill, for one, perennially depended on male assistance; and so too did Ellice Hopkins, whose friendship with the philosopher James Hinton marked a turning-point in her life. Equally, we should recall the sympathetic fathers, husbands, and ministers who encouraged countless less-celebrated women. Without such women, the Jane Gibsons and Mary Cryers, we would not have heard so much of the Octavia Hills and Ellice Hopkinses.

With a genius for fund-raising and organization women fundamentally altered the shape and the course of philanthropy, expanding and redirecting it into channels which suited their perceptions of society and its problems. With few exceptions their charitable work was pragmatic rather than theoretical, often dealing with solutions to immediate and individual problems. This, as we have seen, was very much in tune with their training and with their experience in other spheres. Be that as it may, their contribution to philanthropy was of such magnitude that Clara Lucas Balfour could write, as early as 1849, that 'the history of every religious and benevolent society in the civilized world shows the female sex pre-eminent in numbers, zeal, and usefulness, thus attesting the interest women take in Christian labours for the welfare of society'.[1] She intended this exaggeration to bring more women into the temperance crusade, but in the English experience it was not that far from the mark. Remembering the word of that manly poet unhappy at the prospect of a bazaar, one might say that organized philanthropy had become 'womanized' in the nineteenth century. This transformation took place gradually, largely brought about by the 'noisless influence' of sensible, Christian women, who expanded their activities wherever the opposition of men could be overcome.[2] Pioneers like Hannah More, whose ideal of conservative reform coloured so much of the work of female benevolence, would have applauded.

Whatever one thinks of their influence (one writer argued that 'women make the best and worst philanthropists'),[3] it is obvious that we have been dealing with more than a small band of amateurish ladies, who to fill their idle hours passed

[1] Clara Lucas Balfour, *Women and the Temperance Reformation*, London, 1849, p. 6.

[2] [March Phillipps], *My Life and what shall I do with it?*, p. 13.

[3] Hill, *Women in English Life*, ii, p. 230.

out peppermints.[4] A vast number of women, drawn from virtually every level of society, were responding in their various, increasingly sophisticated ways to what many believed to be a national calamity. Charity was the occupation for which women would always be wanted, said W. R. Greg, and 'society would languish or fall into disorder' without their benevolent industry. He added that 'in a healthy state of civilization [philanthropy] would absorb only a moderate number of women, . . . In our disarranged and morbid state, the demand for their services is enormously enhanced,–enhanced, possibly, almost as much as the supply.'[5] His remarks suggest that virtually every woman who did not receive charity was likely to dispense it, and perhaps this is not very far from the truth. Indeed, it would be difficult to find very many women not on charitable relief who did not contribute in some small way to philanthropic enterprise, if only by an occasional subscription or an act of kindness to a neighbour.

Given the number of women active in philanthropic work, the idea of the idle Victorian woman is difficult to sustain. When Louisa Hubbard, reformer and editor of *The Englishwoman's Yearbook*, carried out a statistical survey of women's work in 1893, in collaboration with Angela Burdett-Coutts, she estimated that about 500,000 women laboured 'continuously and semi-professionally' in philanthropy; another 20,000 supported themselves as 'paid officials' in charitable societies.[6] Her figures did not include some 20,000 trained nurses or the 5,000 women in sisterhoods and nunneries who took on work which was essentially philanthropic. And she obviously omitted the great number of women who worked part time for charity, for example most of the nearly 200,000 members of the Mothers' Union,[7] or the tens of thousands of women who collected money and distributed tracts for the missionary and Bible societies. Her

[4] Jones, *Outcast London*, p. 270, note, gives the impression that passing out peppermints typified female attitudes towards the poor, at least 'traditional aristocratic attitudes'.

[5] W. R. Greg, *Literary and Social Judgements*, London, 1869, p. 314.

[6] Louisa M. Hubbard, 'Statistics of Women's Work', *Woman's Mission*, ed. Burdett-Coutts, pp. 361–6.

[7] *The Quarterly Review*, cvc, 1902, p. 211.

figures are impressive, the more so when put in the context of the 1891 census returns on population and employment. Apart from domestic service no other female occupation came close to 500,000. There were only 24,150 women in agricultural labour and 332,784 in cotton manufacture.[8]

It may be argued that many of the women who enlisted in the expanding charitable societies did so because the paid professions were more difficult to enter. For some this must have been the reasoning, but for the majority such a view is likely to be misleading. Philanthropy itself was becoming more professional, as the number of full-time paid women officials shows. But apart from that, we must not under-estimate the number of women for whom charitable work was an end in itself, who placed the highest value on the rewards of unpaid service. Emily Janes, editor and Secretary of the National Union of Women Workers, remarked in 1893 that 'hardly a girl leaves some of our women's colleges—e.g. Cheltenham and Westfield—without interesting herself in some aspect of philanthropy'.[9] These girls were not without other opportunities, which some of them took up; but with a deeply ingrained sense of social obligation they could not envisage a life cut off from some form of social service. It remained an article of faith right through the century that in the performance of good works woman's nature and mission joined in near perfect harmony.

This is not to say that female philanthropy had no influence on the way in which women viewed themselves or their employments. As we have seen, benevolent women found themselves torn between a desire to express their compassion and morality, which pushed them towards a wider world, and their desire to be modest and unassuming, which kept them back. Belief in their special virtues might be a means to emancipation, as they often argued, but it could also be a snare; for female claims to moral authority and a wider social role were intimately tied to their subordinate position in society. There was a suspicion in the back of

[8] *Census of England and Wales. 1891*, iii, pp. xxvi–xxvii. The total female population of England and Wales aged ten and over in 1891 was 11,461,890.

[9] Emily Janes, 'On the Associated Work of Women in Religion and Philanthropy', *Woman's Mission*, ed. Burdett-Coutts, p. 146.

their minds that if they entered the public arena they might lose their special qualities, for it had been said that the so-called female traits were simply a reflection of woman's dependence, the by-products of lives narrowly confined to home and family. Like medieval churchmen who fell back on morality to increase their power in a society dominated by an armed aristocracy, nineteenth-century women exploited the belief in their superior morality to increase their power in a society dominated by men. They could only give up a belief in the distinctions between the sexes at some peril.

Though in some circles the philanthropic activities of women were thought to be 'contrary to propriety and destructive of modesty', they do not appear to have done very much damage to the commonplaces about female nature.[10] Perhaps benevolent women did lose something of their reputation for retirement and modesty, but they more than made up for it in giving fuller expression to the active and compassionate side of their personalities. Society accepted this state of affairs because it had encouraged their work and felt the benefits. Nor did the movement of women into local government and the professions have any dramatic effect on the views about woman's nature. This may have been because women often entered these pursuits via philanthropy. There was little outcry, for example, when the public elected the first female Poor-Law Guardians to office in the late 1870s; they were already familiar and respected figures because of their charitable work; in London most of them had been active in the COS.[11] The successes of women in philanthropy helped to soften the blow when they turned to those employments which would have been the preserve of men in an earlier time.

[10] Isabella M. S. Tod, *On the Education of Girls of the Middle Classes,* London, 1874, pp. 12–13.

[11] See the manuscript Minutes of the Proceedings of the Guardians of the Poor in the Greater London Record Office and the annual District Committee Reports of the COS. The first female school-board members also had close links with philanthropic societies. For a list of them see *Return of all Members of the Board during its existence*, London, 1904. By the end of the nineteenth century there were over 1,000 female Poor-Law Guardians in Great Britain, over 200 female members of school-boards, and about 200 female parish councillors. *Westminster Review*, cl, 1898, p. 249.

The charitable experience of women was a lever which they used to open the doors closed to them in other spheres, for in its variety it was experience applicable to just about every profession in England.[12] Through their extensive contact with charitable organization women increased their interest in government, administration, and the law. Through contact with charity schools they increased their interest in education. Through the system of district visiting they increased their interest in the problems of poverty and the social services. Through their work as hospital, workhouse, and prison visitors they increased their interest in, among other things, medicine and diet. Moreover, as a religion of action philanthropy slowly challenged the complaisancy of women, gave them practical experience and responsibility, and perhaps most importantly, it heightened their self-confidence and self-respect. At the back of their minds, however, was an awareness that if they were to become more useful they would need more knowledge. Philanthropy pointed out the limitations imposed upon women at the same time as it broadened their horizons. It should not come as a surprise that in 1866 women trained in charitable societies were prominent among those who petitioned the House of Commons praying for the enfranchisement of their sex.[13]

The interest of philanthropic women in female suffrage emerged, not inevitably, but quite naturally out of certain of their activities, for as the more penetrating of them argued, there were limits to their freedom of action without political power. These women were more likely to admire Hannah More than Mary Wollstonecraft,[14] and they often cited their contributions to charity as a justification of their right to vote. To them 'political power really does mean active benevolence'.[15] As Frances Power Cobbe put it, political

[12] For a discussion of the activity of women in teaching, nursing, shops, the civil service, and clerical occupations in the late nineteenth century see Lee Holcombe, *Victorian Ladies at Work*, Newton Abbot, 1973.

[13] Helen Blackburn, *Women's Suffrage: A Record of the Women's Suffrage Movement in the British Isles, with Biographical Sketches of Miss Becker*, London, 1902, Chapter IV.

[14] Helen Blackburn saw Hannah More, Mrs Barbauld, and Mrs Trimmer as passing 'the lamp from hand to hand' and raising the social and intellectual position of women. Ibid., pp. 9–10.

[15] *The Westminster Review*, cxxxii, 1889, p. 279.

emancipation was 'a *means*, a very great means, *of doing good*, fulfilling our Social Duty of contributing to the virtue and happiness of mankind'.[16] This view became a commonplace among campaigners for women's rights in the late nineteenth century, and it suggests that the early female suffrage societies were seen by them as another branch of philanthropy; the means were political but the end philanthropic, the spreading of that 'quality of sympathy and insight which are peculiarly female'.[17] Mrs Fawcett was explicit on this point: 'I advocate the extension of the franchise to women because I wish to strengthen true womanliness in woman, and because I want to see the womanly and domestic side of things weigh more and count for more in all public concerns.'[18] In the nineteenth century one of the strongest arguments in favour of votes for women was the belief that women were *different* from men.[19]

The nineteenth-century female suffrage societies were run along lines established by philanthropic institutions by women who drew on a vast range of charitable experience, not least in administration. The skills required of political association were, after all, the same as those of organized charity: the committee work, the fund-raising and bookkeeping, the public relations, and the lecture tours. No less important were the connections which linked various committee-women around the country. The speed with which the female suffrage societies emerged was a tribute to these connections. Formed almost simultaneously with the London National Society for Women's Suffrage were other female suffrage societies in Birmingham, Bristol, Manchester, Edinburgh, Belfast, and Dublin; and before long over forty smaller societies sprang up in towns around Britain.[20] Without the administrative know-how and social

[16] Frances Power Cobbe, *The Duties of Women*, London, 1881, p. 152.

[17] Butler, ed. *Woman's Work and Woman's Culture*, pp. 278–9. Brian Harrison includes the National Society for Women's Suffrage in his list of moral reform organizations and, as has been noted, discusses the connections between these varied, but related institutions, 'State Intervention and Moral Reform in nineteenth-century England', *Pressure from Without*, p. 319.

[18] Fawcett, *Home and Politics*, p. 8.

[19] *Opinions of Women on Women's Suffrage*, p. 52.

[20] M. G. Fawcett, 'The Women's Suffrage Movement', *The Woman Question*

contacts gained in charitable institutions many of them might well have never been started. And once started other female philanthropic societies lent official support. Lady Henry Somerset's powerful British Women's Temperance Association, for example, set up a political department and announced that votes for women were essential if there was to be temperance reform.[21]

The campaign for women's suffrage drew heavily on the traditions of female philanthropy, but benevolent women did not invariably support the female franchise.[22] They were also prominent among the anti-suffragists, and their charitable role was in tune with the anti-suffrage view of the world.[23] Even astute reformers like Octavia Hill and Angela Burdett-Coutts remained among the ranks of the unconverted. They were chary of hazarding the quiet, domestic influence of their sex for the uncertain rewards of politics. Nor did they wish to tar their respective charitable campaigns by an association with women's suffrage, which aroused such hostilities. Many benevolent women simply did not wish to waste time which they believed could be better spent on philanthropic work; and others, as that shrewd 'old maid' L. F. March Phillipps put it, were 'supinely indifferent to women's rights and stupidly unconscious of woman's wrongs'.[24]

There is no easy formula for determining whether a woman would have been likely to support or oppose the female franchise. The decision was largely a personal one. It

in Europe, ed. Stanton, p. 8. For a list of these local committees with details of their officials see *The National Society for Women's Suffrage: First Report of the Executive Committee*, London, 1872.

[21] *Minutes of the Third Biennial Convention of the World's Woman's Christian Temperance Union*, London, 1893, p. 284.

[22] For a list of philanthropic women in favour of women's suffrage in 1889 see *Declaration in favour of Women's Suffrage, being the Signatures received at the Office of the Central Committee of the National Society for Women's Suffrage*, London, 1889. This list is an extended version of the one published in *The Fortnightly Review*, July 1889.

[23] See Brian Harrison, *Separate Spheres: The Opposition to Women's Suffrage in Britain*, London, 1978, *passim.*

[24] [March Phillipps], *My Life and what shall I do with it?*, p. 9. Perhaps these were the women Dinah Mulock had in mind when she remarked 'who that ever listened for two hours to the verbose confused inanities of a ladies' committee, would immediately go and give his vote for a female House of Commons'?, *A Woman's Thoughts about Women*, p. 5.

is worth making a distinction, however, between a belief in the emancipation of women and support for women's rights.[25] Nineteenth-century definitions of female emancipation tended to leave out politics. And if we take it to mean a freedom from the moral and customary restraints placed on women, or in a pragmatic sense simply the independence to pursue activities and employments outside the home, then charitable women could hardly oppose it; most gave it whole-hearted support. But women's rights, if defined as implying specific political and legal reforms such as votes for women, was a rather more awkward issue for them. Those who were active in the cause of female emancipation, by this definition, were often cool, if not hostile, to the cause of women's rights.

It would be misleading to end this book on the issue of votes for women, as if female philanthropy were simply a preface to the claims of women for political rights. Most nineteenth-century women engaged in more immediate battles with men without attacking them over the vote. For those who had difficulty getting on a charitable committee or who had been refused permission to visit a workhouse or prison because of their sex, votes for women seemed, if not utopian, a distraction. The female philanthropists who did support the extension of the franchise to women were rarely obsessed by the issue, at least in the nineteenth century. Their work for the suffrage societies went hand in hand with their work as moral reformers, institutional visitors, or whatever. They dreamed of the day when men and women would work together as equals in the cause of moral and social reform, dreams nourished by the memory of their predecessors in countless charitable societies who had dealt, often unconsciously, small blow after small blow to the idea of male supremacy. The battle of the sexes in nineteenth-century England was played out in such small arenas.

[25] For a discussion of distinctions between feminism and women's rights in the American context see Barbara Berg, *The Remembered Gate: Origins of American Feminism*, New York, 1978, pp. 4–5.

APPENDIX I*

Societies with Two or More Subscription Lists

Name of society and date of foundation	Dates of the subscription lists	Total number of subscribers	Number of women subscribers	Women as a percentage of all subscribers
Baptist Missionary Society (1792)	1800	364	54	15%
	1825	462	80	17%
	1850	175	40	23%
	1880	196	54	28%
	1900	583	249	43%
Benevolent, or Strangers' Friend Society (1785)	1802	598	81	14%
	1826	2,575	812	32%
	1851	1,496	436	29%
	1870	440	120	27%
British and Foreign Anti-Slavery Society (1839)	1840	301	33	11%
	1854	207	43	21%
	1863	161	38	24%
British and Foreign Bible Society (1804)	1805	1,482	173	12%
	1817	2,712	612	23%
	1854	3,062	930	30%
	1875	2,840	868	31%
	1895	2,431	978	40%
British and Foreign Temperance Society (1831)	1832	289	35	12%
	1835	666	104	16%
British and Foreign Unitarian Association (1825)	1828	184	14	8%
	1845	171	26	15%
	1861	148	26	18%
Church Building Society (1818)	1837	133	22	17%
	1860	1,828	383	21%
Church Lads' Brigade (1891)	1893	171	84	49%
	1898	380	215	57%
Church Missionary Society (1799)	1801	118	14	12%
	1823	1,868	535	29%
	1855	997	330	33%
	1870	909	321	35%
	1900	1,064	522	49%

Name of society and date of foundation	Dates of the subscription lists	Total number of subscribers	Number of women subscribers	Women as a percentage of all subscribers
Church of England Education Society (1853)	1854	1,402	139	10%
	1858	2,130	320	15%
Contagious Fever Institution (1801)	1803	448	35	8%
	1831	547	67	12%
Country Towns Mission (1837)	1849	453	158	35%
	1857	594	217	37%
Foreign Aid Society (1840)	1841	217	97	45%
	1853	375	162	43%
Guardian Society (1812)	1815	465	126	27%
	1831	835	228	27%
	1853	216	59	27%
Hibernian Society (1806)	1808	300	19	6%
	1822	901	128	14%
London City Mission (1835)	1835	90	10	11%
	1838	578	217	38%
	1870	1,134	473	42%
	1901	763	436	57%
London Missionary Society (1795)	1820**	776	131	17%
	1850	283	68	24%
	1880	83	25	30%
	1900	103	45	44%
London Society for the Encouragement of Faithful Female Servants (1813)	1814	136	40	29%
	1820	483	272	56%
Lord's-Day Observance Society (1831)	1833	256	28	11%
	1853	717	193	27%
	1876	551	181	33%
	1900	278	108	39%
Lying-in Charity (1757)	1816	1,260	360	29%
	1833	1,005	287	29%
Manchester Friends' Institute (1858)	1858	105	13	12%
	1890	112	29	26%
Mendicity Society (1818)	1818	1,380	250	18%
	1830	3,375	840	25%
	1864	2,082	501	24%
	1895	879	298	34%
Midnight Meeting Movement (1859)	1861	511	182	36%
	1892	385	133	35%

Name of society and date of foundation	Dates of the subscription lists	Total number of subscribers	Number of women subscribers	Women as a percentage of all subscribers
National Society for Promoting the Education of the Poor (1811)	1812	1,793	212	12%
	1831	2,321	283	12%
	1847	2,461	545	22%
Naval and Military Bible Society (1780)	1812	290	31	11%
	1840	780	200	26%
	1883	718	205	29%
Peace Society (1816)	1817	265	23	9%
	1895	1,428	258	18%
Philanthropic Society (1788)	1816	1,215	275	23%
	1848	844	121	14%
Prayer-Book and Homily Society (1812)	1814	529	55	10%
	1830	1,069	216	20%
	1856	948	225	24%
Pure Literature Society (1854)	1874	963	360	37%
	1885	1,048	453	43%
Reformatory and Refuge Union (1856)	1857	220	22	10%
	1875	1,121	331	30%
	1900	1,180	528	45%
Refuge for the Destitute, Lambeth (1804)	1806	342	49	14%
	1808	580	108	19%
Religious Tract Society (1799)	1801	390	41	11%
	1820	1,362	232	17%
	1840	3,630	1,270	35%
	1860	5,242	1,862	36%
	1880	5,983	2,465	41%
Royal Humane Society (1774)	1827	942	53	6%
	1850	1,110	120	11%
Royal Jennerian Society (1803)	1803	409	29	7%
	1828	575	46	8%
Royal Lancastrian Institution (1808), became the British and Foreign School Society (1814)	1812	606	35	6%
	1815	554	59	11%
	1832	383	75	20%
Royal Society for the Prevention of Cruelty to Animals (1824)	1832***	165	72	44%
	1850	651	325	50%
	1875	2,053	1,081	53%
	1900	2,422	1,673	69%
Saint Ann's Charity School Society (1702) reorganized in 1795	1808	625	21	3%
	1830	2,010	350	17%
	1865	3,680	1,116	30%

Name of society and date of foundation	Dates of the subscription lists	Total number of subscribers	Number of women subscribers	Women as a percentage of all subscribers
School for the Indigent Blind (1799)	1808****	1,674	400	24%
	1816	444	126	28%
Scripture Readers' Association (1844)	1845	390	94	24%
	1851	821	327	40%
	1895	526	225	43%
Society for Bettering the Condition of of the Poor (1796)	1798	246	33	13%
	1805	598	183	31%
Society for Promoting Christian Knowledge (1699)	1790	246	25	10%
	1820	919	193	21%
	1845	15,197	1,956	13%
	1871	11,247	1,547	14%
	1900	9,095	2,740	30%
Society for Promoting Christianity amongst the Jews (1808)	1809	138	16	12%
	1840	815	325	40%
	1870	814	372	46%
	1900	434	254	59%
Society for Promoting the Employment of Additional Curates in Populous Places (1837)	1840	838	131	16%
	1853	1,727	455	26%
Society for Superseding the Necessity of Climbing Boys (1803)	1818	693	152	22%
	1839	150	43	29%
Society for the Conversion and Religious Instruction and Education of the Negro Slaves in the British West India Islands (1820)	1823	209	11	5%
	1829	285	42	15%
Society of Stewards and Subscribers for Maintaining and Educating Poor Orphans of the Clergy (1749)	1790	570	95	17%
	1853	1,000	251	25%
Thames Church Mission Society (1844)	1846	367	72	20%
	1884	2,007	1,168	58%
Trinitarian Bible Society (1831)	1832	437	143	33%
	1885	303	161	53%

Name of society and date of foundation	Dates of the subscription lists	Total number of subscribers	Number of women subscribers	Women as a percentage of all subscribers
United Kingdom Alliance (1853)	1853	56	0	0%
	1864	1,338	98	7%
	1900	2,410	308	13%
York Institution for Persons affected with Disorders of the Mind (1792)	1801	27	3	11%
	1826	431	80	19%

*Subscription lists, though normally limited to name, date, sex, and the amount of each individual's contribution, are reasonably reliable. Dependent on public support, charities were very conscious of the need to keep financial and personal details in good order. Secretaries and accountants made every effort to provide accurate information, and in their accounts they often included addenda and asked readers to correct mistakes and to mention omissions. One does, however, frequently encounter lists with anonymous donors or collections from churches or other societies. The ladies' societies have been counted as female subscriptions; the lists of anonymous contributions and church collections have been omitted. In the case of joint subscriptions from husband and wife every other one has been counted as female. In some institutions, the Methodist Missionary Society for example, the format was too complicated to make a study of female subscriptions worthwhile. In others, some of the Quaker charities for example, initials without the prefix Miss, Mrs, or Mr were too commonly used. The date given for the subscription list refers to the last year in which subscriptions are included. If the list contains subscribers from April 1849 to April 1850, for example, the year 1850 was used. I apologize in advance for any careless mistakes in my counting, but I am satisfied that these would not significantly alter my percentages. I have abbreviated several of the titles. See the Select Bibliography for the various reports, proceedings, etc. of the charities listed in this and the following appendices.

**The London Missionary Society admitted women to its committee in 1891. Only the London subscribers have been counted in the lists cited here, which accounts for their low numbers.

***The RSPCA admitted women to its General Council in 1896.

****The School for the Indigent Blind was not, as these figures suggest, losing its support. The figure for 1808 is for subscribers from 1799 to 1808. The figure for 1816 is for 1816 only. In a few of the other societies as well the lists include back subscribers. As women were increasing in number in most of the charities, the effect of this is to diminish the percentage of women for the latest date.

APPENDIX II

Societies with One Subscription List

Name of society and date of foundation	Date of the subscription list	Total number of subscribers	Number of women subscribers	Women as a percentage of all subscribers
Aborigines' Protection Society (1837)	1840	147	21	14%
African Institution (1806)	1820	500	47	9%
Aged Female Society, Sheffield (1811)	1827	331	261	79%
Alexandra Orphanage for Infants (1864)	1865	570	221	39%
Associate Institution for Improving and Enforcing the Laws for the Protection of Women (1843)	1846	904	193	21%
Associated Catholic Charities (1811)	1836	263	53	20%
Asylum, or House of Refuge, Lambeth (1758)	1808	585	136	23%
British Lying-in Hospital for Poor Married Women (1749)	1805	217	95	44%
Children's Friend Society (1830)	1837	2,588	1,001	39%
Christian Tract Society (1809)	1813	321	69	22%
Church of England Book-Hawking Union (1858)	1867	159	26	16%
City of London Lying-in Hospital (1750)	1816	409	101	25%

Name of society and date of foundation	Date of the subscription list	Total number of subscribers	Number of women subscribers	Women as a percentage of all subscribers
City of London Truss Society (1807)	1821	1,203	55	5%
Colonial Missionary Society (1836)	1853	114	26	23%
Continental Society, for the Diffusion of Religious Knowledge (1818)	1825	896	196	22%
Forlorn Female's Fund of Mercy (1812)	1812	85	27	32%
General Society for Promoting District Visiting (1828)	1832	226	67	30%
Governesses' Benevolent Institution (1841)	1843	623	383	61%
Home Missionary Society (1819)	1853	158	41	26%
Infant Asylum, for the Preserving of the Lives, of Children of Hired Wet-Nurses (unknown)	1799	291	178	61%
Infirmary for Asthma, Consumption, and other Diseases of the Lungs (1814)	1818	259	44	17%
Institution for Rendering Assistance to Shipwrecked Mariners (1808)	1809	54	1	2%
Institution for the Employment of Needlewomen (1860)	1864	120	90	75%
Institution for the Relief of the Poor of the City of London (1799)	1800	1,011	42	4%

Name of society and date of foundation	Date of the subscription list	Total number of subscribers	Number of women subscribers	Women as a percentage of all subscribers
Irish Evangelical Society (1814)	1853*	110	17	15%
Jewish Association for the Diffusion of Religious Knowledge (1860)	1865	200	17	9%
Labourers' Friend Society (1831)	1831	420	16	4%
Liverpool Asylum for Orphan Boys (1850)	1851	368	26	7%
Liverpool Female Orphan Asylum (1840)	1851	895	362	40%
Liverpool Unitarian Fellowship Fund Society (1819)	1823	115	40	35%
Loan Society (1815)	1817	97	63	65%
Lock Asylum for Penitent Females (1787)	1796	154	56	36%
London Female Penitentiary, Pentonville (1807)	1841	834	301	36%
London Society for the Protection of Young Females (1835)	1839	859	146	17%
Maritime Penitent Female Refuge (1829)	1839	649	233	36%
Metropolitan Female Asylum (1829)	1830	71	20	28%
National Sunday League (1855)	1876	233	33	14%
National Temperance Society (1842)	1844	218	27	12%
National Truss Society (1786)	1842	152	5	3%

Name of society and date of foundation	Date of the subscription list	Total number of subscribers	Number of women subscribers	Women as a percentage of all subscribers
New British and Foreign Society for the Suppression of Intemperance (1835)	1840	87	8	9%
Philo-Judaean Society (1826)	1827	60	15	25%
Port of London Society for Promoting Religion among Seamen (1818)	1821	586	34	6%
Ragged Church and Chapel Union (1854)	1857	139	48	35%
Refuge for the Destitute, Hackney (1804)	1818	1,851	311	17%
St. Patrick's Charity (1803)	1838	259	118	46%
Servants' Institution (1834)	1835	154	10	6%
Society for Supplying Home Teachers and Books in Moon's Type (1856)	1859	113	53	47%
Society for the Abolition of Capital Punishment (1828)	1864	250	27	11%
Society for the Diffusion of Knowledge upon Punishment of Death (1808)	1811	288	24	8%
Society for the Discharge and Relief of Persons Imprisoned for Small Debts (1772)	1802	1,300	222	17%
Society for the Extinction of the Slave Trade (1840)	1840	355	27	8%
Society for the Improvement of Prison Discipline and for the Reformation of Juvenile Offenders (1818)	1824	600	72	12%

Name of society and date of foundation	Date of the subscription list	Total number of subscribers	Number of women subscribers	Women as a percentage of all subscribers
Society for the Propagation of the Gospel in Foreign Parts (1701)	1821**	777	113	15%
Society for the Relief of Distress (1860)	1897	255	90	35%
Society for the Relief of Distressed Widows (1823)	1829	913	584	64%
Society for the Relief of the Destitute Sick (1785)	1817	366	111	30%
Society for the Relief of Widows and Orphans of Medical Men (1788)	1818	183	14	8%
Society for the Support and Encouragement of Sunday Schools (1785)	1811	543	50	9%
Society for the Suppression of Vice (1802)	1803	564	177	31%
Society for Visiting and Relieving the Sick Poor, Liverpool (1816)	1819	54	15	28%
Society of Friends of Foreigners in Distress (1807)	1814	906	32	4%
Society of Unitarian Christians, for Promoting Christian Knowledge (1791)	1817	259	20	8%
Training Institution for Nurses, for Hospitals, Families, and the Poor (1848)	1850	357	161	45%
Working Men's Educational Union (1852)	1852	451	40	9%

Name of society and date of foundation	Date of the subscription list	Total number of subscribers	Number of women subscribers	Women as a percentage of all subscribers
Working Men's Lord's Day Rest Association (1857)	1866	967	331	34%

*Life subscribers only.

**Earlier subscription lists are available for this society but because of their format it is impossible to determine the percentage of women in them with any precision.

APPENDIX III

Societies with Mixed Managing Committees of Men and Women

Name of society and date of foundation	Date of the subscription list	Total number of subscribers	Number of women subscribers	Women as a percentage of all subscribers
Charity Organisation Society (1869)	1870	108	25	23%
	1880	1,070	260	24%
	1890	1,551	499	32%
	1900	1,878	632	34%
Essex Hall Temperance Society (1893)	1893	128	39	30%
	1899	125	55	44%
Ethical Society (1886)	1887	32	11	34%
	1896	205	92	45%
Ladies' Charity; or Institution for the Relief of Poor Married Women in Child-bed, Liverpool (1796)	1807	514	468	91%
Metropolitan and National Nursing Association for Providing Trained Nurses for the Sick Poor (1876)	1876	380	164	43%
	1887	255	136	53%
Mission to Women Discharged from Prison (1866)	1867	151	114	76%
Moral Reform Union (1882)	1883	88	68	77%
National Vigilance Association (1885)	1886	228	102	45%
Society for Promoting the Employment of Women (1859)	1867	178	142	80%

Name of society and date of foundation	Date of the subscription list	Total number of subscribers	Number of women subscribers	Women as a percentage of all subscribers
Victoria Street Society for the Protection of Animals from Vivisection (1875)	1895*	229	151	66%
Vigilance Association (1871)	1882	174	114	66%
Women's Protective and Provident League (1874)	1875	54	36	67%
	1889	89	67	75%
Workhouse Visiting Society (1858)	1860**	133	109	82%
	1864	206	174	84%

*These figures are for life members only.

**These figures come from the *Journal of the Workhouse Visiting Society*.

APPENDIX IV

Societies Managed by Women

Name of society and date of foundation	Date of the subscription list	Total number of subscribers	Number of women subscribers	Women as a percentage of all subscribers
British and Irish Ladies' Society (1822)	1828	324	303	94%
Clothing Society for the Benefit of Poor Pious Clergymen of the Established Church and their Families (1820)	1831	256	185	73%
Friendly Female Society, for the Relief of Poor, Infirm, Aged Widows, and Single Women (1802)	1803	500	442	88%
Institution for Nursing Sisters (1840)	1848	236	172	73%
Invalid Asylum for Respectable Females in London and its Vicinity (1825)	1831	400	300	75%
Ladies' Association for Calne, . . . in aid of the Cause of Negro Emancipation (unknown)	1827	112	110	98%
Ladies' Benevolent Society, Liverpool (1810)	1814	262	211	81%
Ladies' Charity School of St. Sepulchre (1702)	1805	325	270	83%
Ladies' Royal Benevolent Society (1812)	1817	980	805	82%

Name of society and date of foundation	Date of the subscription list	Total number of subscribers	Number of women subscribers	Women as a percentage of all subscribers
Liverpool Dorcas and Spinning Society (1818)	1827	134	99	74%
Society of Charitable Sisters (1814)	1841	161	88	55%
Southwark Female Society, for the Relief of Sickness and Extreme Want (1813)	1827	875	663	76%
Women's Vegetarian Union (1895)	1896	206	198	96%

APPENDIX V

The Contribution of Women in Financial Terms: A Sample

Name of Society	Date of the subscription list	Numerical percentage of women subscribers (see appendices I and II)	Total financial contribution of subscribers	Total female financial contribution	Percentage of female financial contribution
Aborigines' Protection Society	1840	14%	£286	£79	28%
Church Lads' Brigade	1893	49%	£323	£129	40%
Church Missionary Society	1801	12%	£178	£15	8%
Colonial Missionary Society	1853	23%	£181	£25	14%
Home Missionary Society	1853	26%	£349	£60	17%
Institution for the Employment of Needlewomen	1864	75%	£208	£149	72%
Irish Evangelical Society	1853	15%	£4,830	£1,868	39%

Jewish Association for the Diffusion of Religious Knowledge	1865	9%	£187	£24	13%
Labourers' Friend Society	1831	4%	£294	£25	9%
Metropolitan Female Asylum	1830	28%	£120	£36	30%
National Temperance Society	1844	12%	£821	£88	11%
Philanthropic Society	1816	23%	£1,854	£402	22%
Prayer-Book and Homily Society	1814	10%	£670	£77	11%
Ragged Church and Chapel Union	1857	35%	£471	£182	39%
Reformatory and Refuge Union	1857	10%	£935	£67	7%
Royal Jennerian Society	1803	7%	£2,748	£258	9%
Society for Promoting Christianity amongst the Jews	1809	12%	£342	£37	11%

Name of Society	Date of the subscription list	Numerical percentage of women subscribers (see appendices I and II)	Total financial contribution of subscribers	Total female financial contribution	Percentage of female financial contribution
Society for Supplying Home Teachers and Books in Moon's Type	1859	47%	£335	£193	58%
Society for the Abolition of Capital Punishment	1864	11%	£447	£30	7%
Society for the Diffusion of Knowledge upon Punishment of Death	1811	8%	£493	£33	7%

APPENDIX VI

Legacies

Name of Society	Years examined	Total number of legators	Number of women legators	Women as a percentage	Total amount of legacies	Amount left by women	Women as a percentage
Baptist Missionary Society	1895–1900	134	75	56%	£30,949	£24,344	79%
Benevolent, or Strangers' Friend Society	1801–26	50	18	36%	£3,291	£731	22%
British and Foreign Bible Society	1870–5	587	291	50%	£133,956	£59,967	45%
Church Missionary Society	1895–1900	511	340	67%	£146,167	£90,961	62%
Colonial Missionary Society	1836–53	32	16	50%	£2,665	£1,177	44%
Guardian Society	1816–76	34	11	32%	£3,499	£600	17%
Home Missionary Society	1852–3	12	6	50%	£770	£325	42%

Name of Society	Years examined	Total number of legators	Number of women legators	Women as a percentage	Total amount of legacies	Amount left by women	Women as a percentage
Irish Evangelical Society	1817–53	66	29	44%	£6,904	£1,374	20%
London City Mission	1865–70	133	88	66%	£27,115	£18,837	69%
London Missionary Society	1875–80	217	109	50%	£39,813	£14,627	37%
Lord's-Day Observance Society	1844–76	39	23	59%	£6,312	£3,652	58%
Methodist Missionary Society	1871–6	178	84	47%	£51,849	£22,162	43%
National Refuges for Homeless and Destitute Children	1853–85	167	80	48%	£54,475	£15,470	28%
National Society for Promoting the Education of the Poor	1815–68	143	44	31%	£71,560	£13,474	19%
Naval and Military Bible Society	1820–82	90	44	49%	£13,368	£5,909	44%

Philanthropic Society	1838–48	31	15	49%	£9,910	£3,310	33%
Prayer-Book and Homily Society	1817–56	86	28	33%	£8,731	£2,982	34%
Reformatory and Refuge Union	1861–1900	79	30	38%	£32,376	£13,945	43%
Religious Tract Society	1855–60	138	63	46%	£15,183	£8,102	53%
Royal Society for the Prevention of Cruelty to Animals	1830–1900	727	473	65%	£301,658	£194,115	64%
Scripture Readers' Association	1851–95	178	112	63%	£30,146	£17,811	59%
Society for Promoting Christian Knowledge	1840–50	106	46	43%	£46,684	£15,826	34%
Society for Promoting Christianity amongst the Jews	1895–1900	272	200	74%	£42,761	£23,070	54%

Name of Society	Years examined	Total number of legators	Number of women legators	Women as a percentage	Total amount of legacies	Amount left by women	Women as a percentage
Society for Promoting the Employment of Additional Curates in Populous Places	1840–53	41	22	54%	£14,323	£7,844	55%
Society for the Propagation of the Gospel in Foreign Parts	1840	21	12	57%	£5,985	£4,258	71%
Society of Friends of Foreigners in Distress	1806–66	34	4	12%	£6,706	£548	8%
Thames Church Mission Society	1852–85	29	14	48%	£3,590	£1,109	31%
Trinitarian Bible Society	1832–85	101	72	71%	£23,613	£20,079	85%
United Brethren Missionary Society	1876–86	41	23	56%	£4,199	£2,959	70%
Totals		4,277	2,372	55%	£1,138,558	£589,568	52%

Select Bibliography*

The place of publication is London unless otherwise stated.

Arranged as follows:

*I have provided a somewhat fuller bibliography than is usual because many of the sources which have informed my work are little known. This is particularly true of the autobiographical remains, which historians of nineteenth-century women commonly believe to be in short supply. I have thus included all of the women's memoirs etc. that I have looked at, most of which did not find their way into the footnotes. For the assiduous student of the British Library Catalogue there are many more yet to be found. Several others, not listed below, are cited in S. Barbara Kanner's useful bibliography in *A Widening Sphere*, ed. Vicinus, pp. 219–21. For historians of children I have listed the various children's periodicals that I found; again they are not all cited in the text. Like the women's memoirs, most of these sources are in the British Library.

I. Manuscripts

Bodleian Library
Ladies' Committee Minutes (1825–37), Church Mission to the Jews
Montagu Manuscripts
British and Foreign Bible Society Archives
Home Correspondence

British Library
Additional Manuscripts
18,591
35,152
35,753
Nightingale Papers
Friends Library
Margaret Woods' Journal
Girton College Library, Cambridge
Bodichon Papers
Greater London Record Office
Minute Book of the Executive Committee of the Society for the Relief of Distress
Minutes of the Proceedings of the Guardians of the Poor
Minutes, Southwark Female Society
Ranyard Papers
Methodist Missionary Society Archive
Notes and Transcripts (now in the School of Oriental and African Studies Library)
Public Record Office
HO 20/3
HO 45 OS6840
HO 45 OS8064/1
HO 45 6628/3
HO 45 7370/11
HO 45 9511/17216
HO 45 9511/17273A
HO 45 9546/59343G
HO 45 9963/X14891
Records of the Prerogative Court of Canterbury, Prob 8 and 11.
School of Oriental and African Studies Library
London Missionary Society Board Minutes

II. Unpublished Theses

Dowling, W. C. 'The Ladies' Sanitary Association and the Origins of the Health Visiting Service' (London M. A., 1963).

Martin, R. H. 'The Pan-Evangelical Impulse in Britain 1795–1830: with special reference to four London Societies' (Oxford D. Phil., 1974).

Yeo, E. M. 'Social Science and Social Change: A Social History of some Aspects of Social Science and Social Investigation in Britain 1830–1890' (Sussex Ph. D., 1972).

III. Parliamentary Papers

Annual Reports of the Poor Law Board, 1836–71.

Annual Reports of the Local Government Board, 1872–1900.

Reports from Committees. Report from Select Committee on the Education of the Lower Orders of the Metropolis, 1816, iv.

Reports from Commissioners. Reports of the Inspectors of Prisons of Great Britain, 1836, xxxv.

Reports from Commissioners. Second Report of the Inspectors of Prisons, 1837, xxxii.

Reports from Committees. Report from the Select Committee of the House of Lords on Poor Law Relief (England), 1861, ix.

Accounts and Papers. Judicial Statistics, 1864, lvii.

Reports from Commissioners. Report of the Royal Commission upon the Administration and Operation of the Contagious Diseases Acts, 1871, xix.

Accounts and Papers. Copy of Annual Report, for 1875, of Captain Harris, Assistant Commissioner of Police of the Metropolis, on the Operation of the Contagious Diseases Acts, 1876, lxi.

Reports from Committees. Report from the Select Committee of the House of Lords on the Law relating to the Protection of Young Girls, 1882, xiii.

Reports from Committees. Report from the Select Committee on Contagious Diseases Acts, 1882, ix.

Reports from Committees. Report from the Select Committee of the House of Lords on Poor Law Relief, 1888, xv.

Reports from Commissioners. Thirty-Fifth Report, for the year 1891, of the Inspector appointed, . . . to visit the Certified Reformatory and Industrial Schools of Great Britain, 1892, xliii.

Reports from Commissioners. Report from the Departmental Committee on Prisons, minutes of evidence, 1895, lvi.

Reports from Commissioners. Eighteenth Report of the Commissioner of Prisons, 1895, lvi.

Reports from Commissioners. Report of the Prison Commissioners, 1898, xlvii.

IV. Reports and Proceedings of Charitable Societies

Annual Reports:

The Baptist Missionary Society
The Benevolent, or Strangers' Friend Society
The British and Foreign Anti-Slavery Society
The British and Foreign Bible Society
The British and Foreign Temperance Society
The British and Foreign Unitarian Association
The British and Irish Ladies' Society
The Church Lads' Brigade
The Church Missionary Society for Africa and the East
The Church of England Education Society
The Church Pastoral-Aid Society
The Essex Hall Temperance Association
The Ethical Society
The Foreign-Aid Society for Promoting the Objects of the European Sociétés Évangéliques
The Friendly Female Society, for the Relief of Poor, Infirm, Aged Widows, and Single Women

The Guardian Society for providing Temporary Asylums for Prostitutes
The Hibernian Society
The Institution for the Cure and Prevention of Contagious Fever in the Metropolis (the London Fever Hospital)
The London City Mission
The London Female Penitentiary
The London Missionary Society
The London Society for Promoting Christianity amongst the Jews
The London Society for the Encouragement of Faithful Female Servants
The Lying-in Charity (later the Royal Maternity Charity for delivering Poor Married Women at their own Habitations)
The Manchester Friends' Institute
The Metropolitan and National Nursing Association for providing Trained Nurses for the Sick Poor
The Midnight Meeting Movement for the Recovery of Fallen Women
The Moral Reform Union
The National Society for Promoting the Education of the Poor in the Principles of the Established Church
The National Society for the Prevention of Cruelty to Children
The National Society for Women's Suffrage
The National Vigilance Association
The Naval and Military Bible Society
The Philanthropic Society
The Prayer-Book and Homily Society
The Pure Literature Society
The Reformatory and Refuge Union
The Refuge for the Destitute, Cuper's Bridge, Lambeth
The Religious Tract Society
The Retreat for Persons afflicted with Disorders of the Mind, York
The Royal Humane Society for the Recovery of Persons apparently Drowned
The Royal Jennerian Society for the Extermination of the Small-Pox
The Royal Lancastrian Institute for the Education of the Poor (later the British and Foreign School Society)
Saint Ann's Schools for Educating, Clothing, and Wholly providing for Offspring of Necessitous Persons (later the Royal Asylum of St. Anne's Society)
The School for the Indigent Blind
The Scripture Readers' Association
The Social Purity Alliance
The Society for Bettering the Condition and Increasing the Comforts of the Poor
The Society for Organising Charitable Relief and Repressing Mendicity (the Charity Organisation Society)
The Society for Promoting Christian Knowledge
The Society for Promoting the Due Observance of the Lord's-Day

The Society for Promoting the Employment of Additional Curates in Populous Places
The Society for Promoting the Employment of Women
The Society for Promoting the Enlargement, Building, and Repairing of Churches and Chapels (the Church Building Society)
The Society for Superseding the Necessity of Climbing Boys
The Society for the Conversion and Religious Instruction and Education of the Negro Slaves in the British West India Islands
The Society for the Prevention of Cruelty to Animals (later the Royal Society for the Prevention of Cruelty to Animals)
The Society for the Promotion of Permanent and Universal Peace (the Peace Society)
The Society for the Propagation of the Gospel in Foreign Parts
The Society for the Suppression of Mendicity (the London Mendicity Society)
The Society of Friends of Foreigners in Distress
The Society of Stewards and Subscribers for Maintaining and Educating Poor Orphans of the Clergy till of Age to be put Apprentice (the Clergy Orphans Society)
The Society of Unitarian Christians for Promoting Christian Knowledge and Virtue
The Thames Church Mission Society
The Town Missionary and Scripture-Readers' Society (later the Country Towns Mission)
The Travellers' Aid Society
The Trinitarian Bible Society
The United Brethren Missionary Society
The United Kingdom Alliance to procure the Total and Immediate Legislative Suppression of the Traffic in all Intoxicating Liquors and Beverages
The Wesleyan Methodist Missionary Society
The Women's Protective and Provident League

Miscellaneous Individual Reports:

Third Annual Report of the Aborigines' Protection Society (London, 1840).
An Account of the Adult Orphan Institution (London, 1831).
Fourteenth Report of the Directors of the African Institution (London, 1820).
Report of the Aged Female Society, 1827 (Sheffield, 1827).
Alexandra Orphanage for Infants, Albert Road, St. John's Ville (London, 1866).
Associate Institution for Improving and Enforcing the Laws for the Protection of Women (London, 1846).
The Associated Catholic Charities for Educating, Clothing, and Apprenticing the Children of Poor Catholics, and Providing for Destitute Orphans ([London], 1836).

Second Annual Report of the Association for the Sale of Work by Ladies of Limited Means ([London, 1859]).

An Abstract from the Account of the Asylum; or House of Refuge, situate in the Parish of Lambeth, (London, 1809).

Bedford Institute, First-Day School and Home Mission Association. Report (London, 1867).

An Account of the British Lying-in Hospital for Married Women, in Brownlow-Street, Long-Acre (London, 1805).

The Children's Aid Society. Its Work and its Aims ([London], 1938).

List of Subscribers to the Children's Friend Society, for the Prevention of Juvenile Vagrancy (London, 1838).

Rules of the Christian Tract Society (London, 1813).

The Tenth Annual Report of the Church of England Book-Hawking Union (London, 1868).

City of London Lying-in Hospital, City Road, for the Reception and Delivery of Poor Pregnant Married Women (London, 1816).

City of London Truss Society, for the Relief of the Ruptured Poor throughout the Kingdom (London, 1822).

Clothing Society for the Benefit of Poor Pious Clergymen of the Established Church and their Families (London, 1832).

Seventeenth Annual Report of the Colonial Missionary Society (London, 1853).

Proceedings of the Continental Society, for the Diffusion of Religious Knowledge over the Continent of Europe (London, 1825).

'These Forty Years'. Being the 40th Annual Report of Dr. Barnardo's Homes (London, [1906]).

The Fourth Annual Report of the General Society for Promoting District Visiting (London, 1832).

Governesses' Benevolent Institution. Report for 1843 (London, 1844).

The Thirty-Fourth Annual Report of the Home Missionary Society (London, 1853).

Infant Asylum, for the preserving of the lives, of children of hired wet-nurses, and others (London, 1799).

Account of the Infirmary for Asthma, Consumption, and other Diseases of the Lungs (London, 1818).

Report of the Institution for Nursing Sisters (London, 1848).

Plan of an Institution, for rendering assistance to Shipwrecked Mariners (London, 1810).

Institution for the Employment of Needlewomen (London, 1864).

Fund of Mercy; or an Institution for the Relief and Employment of Destitute and Forlorn Females (London, 1813).

Report of the State and Progress of the Institution for the Relief of the Poor of the City of London and Parts Adjacent (London, 1800).

Thirty-Ninth Annual Report of the Irish Evangelical Society (London, 1853).

Fifth Report, for 1831, of the Invalid Asylum, for Respectable Females in London and its Vicinity (London, 1832).

Fifth Annual Report of the Jewish Association for the Diffusion of Religious Knowledge (London, 1865).
Proceedings of the Labourers' Friend Society (London, 1832).
The Second Report of the Ladies' Association for Calne, . . . in aid of the Cause of Negro Emancipation (Calne, 1827).
The Fifth Annual Report of the Ladies' Benevolent Society, Liverpool (Liverpool, 1815).
The Eleventh Report of the Ladies' Charity; or Institution for the Relief of Poor Married Women in Child-bed, Liverpool (Liverpool, 1807).
Plan of the Ladies' Charity School of St. Sepulchre, London (London, 1805).
The Ladies' Royal Benevolent Society (Late Dollar) for Visiting, Relieving, and Investigating the Condition of the Poor at their own Habitations (London, 1818).
Extract from an Account of the Ladies' Society for the Education and Employment of the Female Poor (London, 1804).
Report of the Liverpool Asylum for Orphan Boys (Liverpool, 1851).
The First Annual Report of the Liverpool District Provident Society, for the Year 1830 (Liverpool, [1831])
Tenth Report of the Liverpool Dorcas and Spinning Society (Liverpool, 1827).
Report of the Liverpool Female Orphan Asylum (Liverpool, 1851).
The Fifth Annual Report of the Committee of the Liverpool Unitarian Fellowship Fund Society (Liverpool, 1824).
The Second Annual Report of the Loan Society (London, 1817).
An Account of the Institution of the Lock Asylum for the Reception of Penitent Females (London, 1796).
Fourth Report of the London Society for the Protection of Young Females (London, 1839).
Tenth Annual Report of the Maritime Penitent Female Refuge, for affording protection to innocent Females (London, 1839).
Report of the Metropolitan Association for Befriending Young Servants for 1886 (London, 1887).
The Report of the Provisional Committee of the Metropolitan Female Asylum (London, 1830).
Report of Mission to Women Discharged from Prison [London, 1868].
The Forty-First Report of the National Refuges for Homeless and Destitute Children (London, 1886).
Annual Report of the Council of the National Sunday League (London, 1877).
National Temperance Society . . . . Second Report (London, 1844).
National Truss Society for the Relief of the Ruptured Poor (London, [1842]).
The Third Report of the New British and Foreign Temperance Society (London, 1840).
First Report of the Philo-Judaean Society (London, 1827).

Proceedings of the Third Anniversary of the Port of London Society for Promoting Religion among Seamen (London, 1821).
The Fourth Annual Report of the Ragged Church and Chapel Union (London, 1857).
A Short Account of the Refuge for the Destitute, Hackney Road, and Hoxton (London, 1818).
The Eighteenth Annual Report of the Rescue Society (London, 1872).
St. Patrick's Charity, for the Gratuitous Education and Clothing of the Children of Poor Catholics, and Asylum for Female Orphans ([London], 1838).
The Servants' Institution (London, 1835).
Report of the Society for Relief of the Destitute Sick (London, 1817).
Society for Relief of Widows and Orphans of Medical Men, in London and its Vicinity (London, 1818).
Third Report of the Society for Supplying Home Teachers and Books in Moon's Type, to enable the Blind to read the Scriptures (London, 1859).
Report of the Society for the Abolition of Capital Punishment (London, 1865).
An Account of the Origin and Object of the Society for the Diffusion of Knowledge upon the Punishment of Death, and the Improvement of Prison Discipline (London, 1812).
An Account of the Rise, Progress, and Present State, of the Society for the Discharge and Relief of Persons Imprisoned for Small Debts (London, 1802).
Proceedings at the First Public Meeting of the Society for the Extinction of the Slave Trade, and for the Civilization of Africa (London, 1840).
The Sixth Report of the Committee of the Society for the Improvement of Prison Discipline, and for the Reformation of Juvenile Offenders (London, 1824).
Thirty-Seventh Annual Report of the Society for the Relief of Distress for the Year 1897 (London, 1897).
The Sixth Annual Report of the Society for the Relief of Distressed Widows (London, 1830).
Plan of a Society, established in London, for the Support and Encouragement of Sunday-Schools (Bermondsey, 1812).
Part the First, of an Address to the Public, from the Society for the Suppression of Vice (London, 1803).
Second Report of the Society for Visiting and Relieving the Sick Poor (Liverpool, [1820]).
Report of the Society of Charitable Sisters, for 1841 (London, 1842).
Articles and Regulations proposed for the Society of Universal Good-Will in London, or Elsewhere, 1789 (Norwich, 1789).
Report of the Proceedings at the Horns Tavern, Kennington, . . . when an Auxiliary Bible Society was established, for Southwark and its Vicinity (Southwark, 1812).

The Fourteenth Annual Report of the Southwark Female Society, for the Relief of Sickness and Extreme Want (London, 1827).
Rules, Bye-laws, and Regulations, of the Training Institution for Nurses, for Hospitals, Families and the Poor (London, 1850).
Twentieth Annual Report of the Victoria Street Society for the Protection of Animals from Vivisection (London, 1895).
Twelfth Annual Report of the Vigilance Association for the Defence of Personal Rights, and for the Amendment of the Law (London, 1883).
Report of the Visitation of Females at their own Homes, in the City of Westminster (London, 1854).
White Cross League Church of England Society Twelfth Annual Report (London, 1895).
First Annual Report of the Women's Total Abstinence Union for the Year 1893-4 (London, [1894]).
First Annual Report of the Women's Vegetarian Union (London, 1896).
The First Annual Report of the Working men's Educational Union (London, 1853).
The Ninth Annual Report of the Working men's Lord's Day Rest Association (London, 1866).

V. Newspapers and Periodicals etc.

All the World
The Annual Charities Register and Digest
The Anti-Jacobin Review
The Anti-Slavery Reporter
Blackwood's Magazine
The Book and its Mission
The Brighton Herald
Bristol Gazette
Bristol Mercury
British Mothers' Magazine
The British Workwoman Out and at Home
The Cambridge Chronicle and University Journal
The Cambridge Independent Press
Chambers's Miscellany of Useful and Entertaining Tracts
The Chelsea News and General Advertiser
Chelsea and Pimlico Advertiser
The Christian Family
Christian Lady's Magazine
The Christian Mother's Magazine
The Christian Observer
The Christian Remembrancer
The Church Missionary Intelligencer
Church Missionary Record
Church Progress. A Monthly Record of Home Mission Work
Churchman's Family Magazine

The Contemporary Review
Cornhill Magazine
The Countess of Huntingdon's New Magazine
Court Magazine and Belle Assemblée
Daily Mail
Daily Telegraph
The Deliverer
East London Advertiser
The East London News
The Edinburgh Review
The Englishwoman
The English Woman's Journal
The Englishwoman's Magazine
The Englishwoman's Review
The Evangelical Magazine
The Fancy Goods and Toy Trades Journal
The Female's Friend
The Fortnightly Review
Fraser's Magazine
Friendly Leaves
The Girl of the Period Miscellany
Good Words
The Hammersmith Advertiser
The Hampstead Record
Hansard Parliamentary Debates
The Harbinger
Howitt's Journal
The Illustrated London News
The Indian Female Evangelist
Journal of the Royal Statistical Society
Journal of the Workhouse Visiting Society
The Ladies' Cabinet of Fashion
The Ladies' Penny Gazette
Leeds Mercury
The London Quarterly Review
Macmillan's Magazine
The Magdalen's Friend
Meliora
The Methodist Temperance Magazine
The Missing Link Magazine
Missionary Leaves
The Monthly Packet of Evening Readings
Monthly Record of Christian Labour among Women and Children
Monthly Reporter of the British and Foreign Bible Society
National Anti-Corn-Law League Bazaar Gazette
The Needle. A Magazine of Ornamental Work
The Newcastle Daily Journal
The Nineteenth Century

North British Review
The Philanthropist
Pioneer
Press Bazaar News
Punch
The Quarterly Review
The Reformatory and Refuge Journal
The Royal Lady's Magazine, and Archives of the Court of St. James's
Seeking and Saving
The Shield
The Sketch
The Suffragette
A Threefold Cord
The Times
Transactions of the National Association for the Promotion of Social Science
Vanity Fair
The Victoria Magazine
The War Cry
The Wesleyan Methodist Magazine
The Westminster Review
The Woman's Gazette
Woman's Work in the Great Harvest Field
Work and Leisure
The Englishwoman's Yearbook

VI. Children's Periodicals

The Baptist Children's Magazine and Youth's Missionary Repository
The Bible Class Magazine
The British Juvenile, at Home, at Work, and at Play
Brothers and Sisters. A Quarterly Paper for Children
The Children's Corner
The Children's Hour for the Young of the Fold
The Children's Magazine
Children's Magazine and Missionary Repository
Children's Missionary Magazine
The Children's World
The Church Missionary Juvenile Instructor
The Church of England Sunday Scholar's Magazine
The Girls' and Boys' Penny Magazine
Gleanings for the Young
Golden Sunbeams
The Juvenile
Juvenile Bethel Flag
The Juvenile Christian's Remembrancer
The Juvenile Companion and Sunday School Hive
The Juvenile Forget Me Not

The Juvenile Gleaner
The Juvenile Instructor and Companion
The Juvenile Keepsake
Juvenile Magazine
Juvenile Messenger
Juvenile Messenger of the Presbyterian Church in England
The Juvenile Miscellany of Facts and Fictions
The Juvenile Missionary Herald
The Juvenile Missionary Magazine
The Little Messenger
Little Ones at Home
The Little Standard Bearer
A Monthly Missionary Letter to Sunday Schools
The Morning of Life
The Primitive Methodist Children's Magazine
Primitive Methodist Juvenile Magazine
A Quarterly Juvenile Review
A Quarterly Token for Juvenile Subscribers
Sunday. Reading for the Young
The Wesleyan Juvenile Offering
Youth's Magazine or Evangelical Miscellany

VII. Miscellaneous Printed Primary Sources

Acton, William. *Prostitution*, ed. Peter Fryer (1968).

Amos, Sarah M. 'The Prison Treatment of Women'. *Contemporary Review*, lxxiii (1898), pp. 803–13.

Anderson, Christopher. *The Domestic Constitution; or, the Family Circle the source and test of national stability* (Edinburgh, 1847).

Annals of Banks for Savings (1818).

Anon. *The Bazaar; or Fragments of Mind* (Lancaster, 1831).

Anon. [E. Bickersteth]. *Woman's Service on the Lord's Day* (1861).

Anon. 'Charity as a Portion of the Public Vocation of Women'. *The English Woman's Journal*, iii (1859), pp. 193–6.

Anon. *District Visitors, Deaconesses, and a proposed adaptation, in part, of the Third Order* (1890).

Anon. *The Fancy Fair* (1833).

Anon. [William Atkinson Gardner]. *A Rhythmical Notice of the Anti-Corn-Law League Bazaar* (1845).

Anon. 'A Group of Female Philanthropists'. *London Quarterly Review*, lvii (1881), pp. 49–81.

Anon. 'Helping the Poor', *Meliora*, vii (1864), pp. 237–46.

Anon. *The House of Mercy at Ditchingham* (Oxford and London, 1859).

Anon. [Louisa M. Hubbard]. *A Guide to all Institutions Existing for the Benefit of Women and Children*, 5 Parts (1878–80).

Anon. *Memoirs of an Unappreciated Charity*, from *Good Words* (1879).

Anon. 'The Missing Link and the London Poor'. *Quarterly Review*, cviii (1860), pp. 1–34.

Anon. 'Parochial Mission Women'. *Meliora*, xii (1869), pp. 89–91.

Anon. 'The Progress of Women', *Quarterly Review*, cvc (1902), pp. 201–20.

Anon. 'The Social Position of Woman'. *North British Review*, xiv (1851), pp. 275–89.

Anon. 'Social Science', *Blackwood's Magazine*, xc (1861), pp. 463–78.

Anon. 'Some of the Work in which Women are Deficient'. *The English Woman's Journal*, iii (1859), pp. 190–6.

Anon. 'The Vices of the Streets' *Meliora*, i (1858), pp. 70–9.

Anon. *A Visit to the Bazaar* (1818).

Anon. 'What Woman is Fitted for'. *Westminster Review*, cxxvii (1887), pp. 64–75.

Anon. *Woman: as she is, and as she should be*, 2 vols. (1835).

Anon. *Woman, her Character and Vicissitudes* (1845).

Archer, Hannah. *A Scheme for befriending Orphan Pauper Girls* (1861).

Armstrong, J. 'Female Penitentiaries'. *Quarterly Review*, lxxxiii (1848), pp. 359–76.

The Autobiography of a Missionary Box [1896].

The Autobiography of Mark Rutherford, ed. Reuben Shapcott (1881).

Bagehot, Walter. *Physics and Politics* (1872).

Balfour, Clara Lucas. *Women and the Temperance Reformation* (1849).

—— *The Women of Scripture* (1847).

—— *Working Women of the last Half Century: the lesson of their lives* (1854).

Barber, Mary Ann. *Missionary Tales, for little listeners* (1840).

Barlee, Ellen. *A Visit to Lancashire in December, 1862* (1863).

—— Sunshine; or Believing and Rejoicing: A series of home and foreign missionary sketches (1854).

Bayly, Mrs Mary. *Mended Homes, and what repaired them* (1861).

—— *Ragged Homes, and how to mend them* (1860).

Bennett, Arnold. *Anna of the Five Towns* (1902).

Bernard, Thomas. *Pleasure and Pain, 1780–1818*, ed. J. Bernard Baker (1930).

Bevan, F. L. [Mrs Mortimer]. *The Peep of Day: or a Series of the earliest religious instruction the infant mind is capable of receiving* (1870).

Bible Associations Exposed. Being a Review of the Fourth Annual Report, of the Committee of the Henley Bible Society (1818).

Bishop, J. C. *'A Peep into the Past': Brighton in the Olden Time* (Brighton, 1880).

Blackburn, Helen. *Women's Suffrage: A Record of the Women's Suffrage Movement in the British Isles, with Biographical Sketches of Miss Becker* (1902).

Blackmore, Lieut. John. *The London by Moonlight Mission* (1860).

Blake, Joseph. *The Day of Small Things* (Sheffield, 1868).

Blunt, J. J. *The Duties of a Parish Priest* (1856).

Bolton, James. *The Golden Missionary Penny, and other addresses to the young* (1868).

Bolton, James. *Missionary Stick Gatherers: an address to the members of juvenile missionary associations* (1854).
Booth, General William. *In Darkest England and the Way Out* (1890).
Bosanquet, Charles B. P. *London: Some Account of its Growth, Charitable Agencies, and Wants* (1868).
Boucherett, Jessie. *Hints on Self-Help; a book for young women* (1863).
Bremer, Frederika. *England in 1851; or, Sketches of a Tour in England* (Boulogne, 1853).
Briggs, Frederick W. *Chequer Alley* (1866).
Brinckman, Arthur. *Notes on Rescue Work* (1885).
Broadhurst, Thomas. *Advice to Young Ladies on the Improvement of the Mind and the Conduct of Life* [1808].
Brock, Mrs Carey. *Charity Helstone. A tale* (1866).
Browning, Elizabeth Barrett. *Aurora Leigh* (1857).
Burns, Jabez. *The Marriage Gift Book and Bridal Token* (1863).
Butler, Josephine E. *Personal Reminiscences of a Great Crusade* (1896).
—— ed. *Woman's Work and Woman's Culture* (1869).
Cappe, Catharine. *Observations on Charity Schools, Female Friendly Societies, and other Subjects connected with the views of the Ladies' Committee* (York, 1805).
—— *Thoughts on the Desirableness and Utility of Ladies visiting the Female Wards of Hospitals and Lunatic Asylums* (1816).
Carter, The Revd. T. T. *The First Ten Years of the House of Mercy, Clewer* (1861).
—— *Is it well to institute Sisterhoods in the Church of England, for the care of female penitents?* (1851).
—— *Objections to Sisterhoods considered, in a Letter to a Parent* (1853).
The Catholic Directory, Ecclesiastical Register, and Almanac, for . . . 1880 (1880).
Chance, William. *Children under the Poor Law* (London, 1897).
The Charitable Ten Thousand (1896).
Charles, A. O. *The Female Mission to the Fallen* (1860).
Children's Fancy Work: A Guide to Amusement and Occupation for Children [1882].
The Children's Missionary Meeting in Exeter Hall, on Easter Tuesday, 1842 (1842).
A Christian Poet [John Holland]. *The Bazaar; or Money and the Church* (Sheffield [1830?]).
Church Missionary Society Centenary Meeting for Boys and Girls (1899).
A Churchman. *A Letter to the Church Members of the Auxiliary Bible Society, Liverpool* (Liverpool, 1819).
Clayton, Louisa. *The London Medical Mission. What is it Doing?* (1873).
Cobbe, Frances Power. *The Duties of Women* (1881).

Cobbe, Frances Power. *Essays on the Pursuits of Women* (1863).
— 'Social Science Congresses, and Women's Part in Them'. *Macmillan's Magazine*, v (1861), pp. 81-94.
— *Why Women desire the Franchise* [1877].
— 'Workhouse Sketches'. *Macmillan's Magazine*, iii (1861), pp. 448-61.
Collins, Wilkie. *The Moonstone. A Romance*, 3 vols. (1868).
Conybeare, William. *Charity of Poor to Poor, facts collected in South London at the suggestion of the Bishop of Southwark* (1908).
Coote, William Alexander. *A Romance of Philanthropy* (1916).
Correspondence of William Ellery Channing, D. D. and Lucy Aikin, ed. Anna Letitia le Breton (1874).
A Country Lady. *Females of the Present Day, considered as to their influence on Society* (1831).
Crawshay, Rose Mary. *Domestic Service for Gentlewomen: A Record of Experience and Success* (1874).
Declaration in favour of Women's Suffrage, being the Signatures received at the Office of the Central Committee of the National Society for Women's Suffrage (1889).
Dibdin, Thos. Frognall. *Reminiscences of a Literary Life*, 2 Parts (1836).
Dickens, Charles. *American Notes*, 2 vols. (1842).
— *Sketches by 'Boz'*, 2 vols. (1836).
Dolling, Robert R. *Ten Years in a Portsmouth Slum* (1896).
Drummond, Mrs Harriet. *Louisa Moreton; or, children obey your parents in all things* [1871].
— *Lucy Seymour; or, it is more blessed to give than to receive* (Edinburgh, 1847).
Dudley, C. S. *An Analysis of the system of the Bible Society throughout its various parts* (1821).
Elliott, Charlotte. *The Invalid's Hymn-Book* (Dublin, 1834).
Ellis, Emily Glode. 'The Fetish of Charity'. *Westminster Review*, cxxxv (1891), pp. 300-10, 373-84.
Ellis, Mrs Sarah. *The Daughters of England* [1845].
— *The Education of Character; with hints on moral training* (1856).
— *Education of the Heart: woman's best work* (1869).
— *The Wives of England; their relative duties, domestic influence, and social obligations* [1843].
Engels, Friedrich. *The Condition of the Working Class in England*, eds. W. O. Henderson and W. H. Chaloner (Stanford, 1958).
Evans, R. W. *The Bishopric of Souls* (1841).
Everard, George. *Bright and Fair. A Book for young ladies* (1882).
Faber, Frederick William. *Spiritual Conferences* (1859).
Fawcett, Mrs H. *Home and Politics* (1898).
— 'The Women's Suffrage Movement'. *The Woman Question in Europe*, ed. Theodore Stanton (1884), pp. 1-30.
[Female author]. *A Visit to a Religious Bazaar* (1857).
A Few Words to Servants about the Church Penitentiary Association [1854].

Fifty Years' Record of Child-Saving and Reformatory Work (1856–1906) [1906].

The First Forty Years. A Chronicle of the Church of England Waifs & Strays Society 1881–1920 (London, 1922).

Forster, John. *The Life of Charles Dickens* [1879].

Frazer, Catherine. 'The Origin and Progress of "The British Ladies' Society for Promoting the Reformation of Female Prisoners", established by Mrs. Fry in 1821.' *Transactions of the National Association for the Promotion of Social Science*, vi (1862), pp. 495–501.

Freeman, Flora Lucy. *Religious and Social Work amongst Girls* (1901).

Fremantle, W. H. *St. Mary's, Bryanston Square: Pastoral address and report of the charities of the year 1870* [1871].

Garnett, Mrs Elizabeth. *Our Navvies: A Dozen Years Ago and To-day* (1885).

The Gladstone Diaries. iii, 1840–1847, eds. M. R. D. Foot and H. C. G. Matthew (Oxford, 1974).

Goodman, Margaret. *Sisterhoods in the Church of England* (1863).

Gordon, Margaret Maria. *The Double Cure; or, What is a Medical Mission?* (1869).

Graham, Catherine Macaulay. *Letters on Education* (1790).

Grant, James. *Sketches in London* (1838).

Greg, W. R. *Literary and Social Judgements* (1869).

—— 'Prostitution'. *Westminster Review*, liii (1850), pp. 238–68.

Gregory, Robert. *The Difficulties and the Organization of a Poor Metropolitan Parish* (1866).

Greville, Charles C. F. *A Journal of the Reigns of King George IV and King William IV*, 3 vols. (1875).

[E. Grubb?]. *Statistics of Middle-Class Expenditure*, British Library of Political and Economic Science Pamphlet HD6/D267 (1896).

A Guide to Bazaars and Fancy Fairs, the Organisation and Management, with Details of the Various Devices for Extracting Money from the Visitors [1882].

Handbook of Catholic Charitable and Social Works (1905).

Handbook of Catholic Charities, Associations, etc., in Great Britain (1894).

Hemmens, H. L. *Such has been my Life* (1953).

Higgs, Mary. *Glimpses into the Abyss* (1906).

Highmore, Anthony. *Philanthropia Metropolitana* (London, 1822).

—— *Pietas Londinensis*, 2 vols. (1810).

Hill, Florence. *Children of the State: the training of Juvenile Paupers* (1868).

Hill, Georgiana. *Women in English Life from Medieval to Modern Times*, 2 vols. (1896).

Hill, Octavia. *District Visiting* (1877).

—— *Homes of the London Poor* (1875).

—— *Letter accompanying the Account of Donations received for Work amongst the Poor during 1872* (1873).

Hill, Octavia. *Our Common Land (and other short essays)* (1877).
Hilton, Marie. *The Fourth Year of the Crèche* [1875].
Hints on Ladies' Working-Parties, and Supplies for Missionary Stations (1857).
Hoare, The Revd. E. A. *Notable Workers in Humble Life* (1887).
Hopkins, Ellice. *Drawn unto Death* (1884).
— *Grave Moral Questions* (1882).
— *Home Thoughts for Mothers and Mothers' Meetings* (1869).
— *How to Start Preventive Work* (1884).
— *Ladies' Associations for the Care of Friendless Girls* (1878).
— *Notes on Penitentiary Work* (1879).
— *A Plea for the Wider Action of the Church of England in the Prevention of the Degradation of Women* (1879).
— *The Power of Womanhood* (1899).
— *The Present Moral Crisis: an appeal to women* [1886].
— *The Visitation of Dens* (1874).
— *The Visitation of Dens. An Appeal to the Women of England* (1874).
— *The White Cross Army* [1883].
— *The White Cross Army. A statement of the Bishop of Durham's Movement* (1883).
— *Work in Brighton; or, Woman's Mission to Women* (1877).
Howe, W. F. *Twenty-Fourth Annual Edition of the Classified Directory to the Metropolitan Charities for 1899* (1899).
Howson, J. S. 'Deaconesses'. *Quarterly Review*, cviii (1860), pp. 342-87.
— *Deaconesses; or the official help of women in parochial work and in charitable institutions* (1862).
Hubbard, Louisa M. 'Statistics of Women's Work'. *Women's Mission*, ed. Baroness Burdett-Coutts (1893), pp. 361-6.
Hume, A. *Analysis of the Subscribers to the Various Liverpool Charities* (Liverpool, 1855).
Ignota. 'The Part of Women in Local Administration'. *Westminster Review*, cl (1898), pp. 32-46, 248-60.
James, John Angell. *The Family Monitor, or a Help to Domestic Happiness* (Birmingham, 1828).
— *Female Piety: or, the Young Woman's Friend and Guide through Life to Immortality* (1852).
— *Juvenile advantages and obligations in reference to the cause of Christian missions stated and enforced. A Sermon.* (1828).
— *The Mother's Help towards instructing her Children* (1842).
Jameson, Anna. *The Communion of Labour* (1855).
— *Sisters of Charity and the Communion of Labour* (1859).
Janes, Emily. 'On the Associated Work of Women in Religion and Philanthropy'. *Woman's Mission*, ed. Baroness Burdett-Coutts (1893), pp. 131-48.
[Kaye, J. W.]. 'The Employment of Women'. *North British Review*, xxvi (Feb. 1857), pp. 157-82.

Kenney, Martha. *Charity: a poem* (1823).
Ladies at Work. Papers on paid employments for ladies. By experts in the several branches. Introduced by Lady Mary Jeune (1893).
The Ladies' National Temperance Convention of 1876 (1876).
A Lady. *Hints for Lady Workers at Mission Services* (n.d.).
A Lady [Martha Nicol]. *Ismeer; or Smyrna and its British Hospital in 1855* (1856).
A Lady. *The Whole Duty of Woman or, a Guide to the female sex, from the age of sixteen to sixty* (Stourbridge, 1815).
A Lady. *Woman! as Virgin, Wife and Mother* [1838].
Landels, William. *Lessons for Maidens, Wives, and Mothers, from some of the Representative Women of Scripture* (1865).
—— *Woman: Her Position and Power* [1870].
—— *Woman's Sphere and Work, considered in the light of Scripture* (1859).
Lathbury, Bertha. 'Agnosticism and Women'. *The Nineteenth Century*, vii (1880), pp. 619–27.
Layton, W. T. 'Changes in the Wages of Domestic Servants during Fifty Years'. *Journal of the Royal Statistical Society*, lxxi (1908), pp. 515–24.
Lectures to Ladies on Practical Subjects (Cambridge, 1855).
Leigh, Samuel. *New Picture of London* (1830).
Linton, E. Lynn. *The Girl of the Period and Other Social Essays* (1883).
—— *Ourselves. A series of essays on Women* (London and New York, 1870).
Lister, T. H. 'Rights and Conditions of Women'. *Edinburgh Review*, lxxiii (1841), pp. 189–209.
Lloyd, The Revd. Richard. *Strictures on a recent publication entitled 'The Church her own Enemy'; to which are added, a refutation of the arguments contained in the Rev. Edward Cooper's Letter to the Author; and an admonitory address to the female sex* (1819).
Lovett, Richard. *The History of the London Missionary Society, 1795–1895*, 2 vols. (1899).
Low, Sampson. *The Charities of London* (1850).
McKenzie, F. A. *Waste Humanity* (1908–09).
Maddocks, Mrs. *The Female Missionary Advocate* (1827).
[March Phillipps, Lucy F.]. *My Life and what shall I do with it?* (1860).
[Marsh, C. M.]. *Death and Life* (1867).
—— *English Hearts and English Hands* (1858).
Martineau, Harriet. 'Life in the Criminal Class'. *Edinburgh Review*, cxxii (1865), pp. 337–71.
Martyn, Elizabeth. 'Women in Public Life'. *Westminster Review*, cxxii (1889), pp. 278–85.
Mayhew, Henry. *London Labour and the London Poor*, 4 vols. (1861). The author of vol. iv was Bracebridge Hemyng.
Meredith, Mrs Susanna. *A Book about Criminals* (1881).
—— *Wanted, Deaconesses for the Service of the Church* [1872].
Merryweather, Mary. *Experience of Factory Life* (1862).

Mill, John Stuart. *The Subjection of Women* (1869).

Minutes of the Third Biennial Convention of the World's Woman's Christian Temperance Union (1893).

Montague, C. J. *Sixty Years in Waifdom* (1904).

More, Hannah. *Coelebs in search of a Wife*, 2 vols. (1809).

—— *Moral Sketches of Prevailing Opinions and Manners* (1819).

—— *Strictures on the Modern System of Female Education*, 2 vols. (1799).

[Muloch, Dinah Maria]. *A Woman's Thoughts about Women* (1858).

National Anti-Corn-Law League Bazaar, to be held in Covent Garden Theatre, London, May, 1845 [1844].

National Anti-Corn-Law-League Quarter of a Million Fund (Manchester, 1846).

Nichols, James. *A Report of the Principal Speeches delivered on the Sixth Day of October 1813, at the Formation of the Methodist Missionary Society, for the Leeds District* (Leeds, 1813).

[Nightingale, Florence]. *Suggestions for Thought to Searchers after Religious Truth*, 3 vols. (1860).

Norris, H. H. *The Origin, Progress, and Existing Circumstances, of the London Society for Promoting Christianity amongst the Jews* (1825).

Orme, Eliza. 'Our Female Criminals'. *Fortnightly Review*, lxix (1898), pp. 790–6.

Opinions of Women on Women's Suffrage (1879).

Parent-Duchatelet, A. J. B. *De la prostitution dans la ville de Paris* (Paris, 1857).

Parker, Theodore. *A Sermon on the public function of Woman* (Boston, 1853).

Parkes, B. R. 'A Year's Experience in Woman's Work'. *The English Woman's Journal*, vi (1860), pp. 112–21.

A Parson. *My District Visitors* (1891).

Penitentiary Work in the Church of England (1873).

A Physician [Gustave Richelot]. *The Greatest of our Social Evils*, (1857).

The Poor Law System of Elberfeld [1870].

Prentice, Archibald. *History of the Anti-Corn-Law League*, 2 vols. (1853).

[Price, Eleanor C.]. *Schoolboy Morality: an address to mothers* (1886).

A Prison Matron [Frederick William Robinson]. *Female Life in Prison*, 2 vols. (1862).

—— *Prison Characters drawn from life, with suggestions for prison government*, 2 vols. (1866).

[Ranyard, Ellen]. *Life Work; or the Link and the Rivet* (1861).

—— *London, and Ten Years Work in It* (1868).

—— *The Missing Link; or Bible-Women in the Homes of the London Poor* (1859).

—— *Nurses for the Needy or Bible-Women in the Homes of the London Poor* (1875).

[Ranyard, Ellen]. *The True Institution of Sisterhood: or a Message and its Messengers* [1862].
[Rathbone, William]. *Social Duties* (1867).
Reaney, Mrs. G. S. *Our Daughters: their lives here and hereafter* (1891).
—— *Waking and Working* (1874).
Reffold, A. E. *The Audacity to Live* [1938].
Reid, Mrs [Marion]. *A Plea for Woman: being a Vindication of the Importance and Extent of her Natural Sphere of Action* (Edinburgh, 1843).
Return of all Members of the Board during its existence (1904).
Ritchie, James Ewing. *The Night Side of London* (1857).
Ruskin, John. *Sesame and Lilies* (1865).
Ryan, Michael. *Prostitution in London* (1839).
Sandford, Mrs Elizabeth. *Woman, in her social and domestic character* (1831).
Sapsworth, L. *The Emancipation of Women* (1913).
Schimmelpenninck, Mary Anne. *Sacred Musings on Manifestations of God to the Soul of Man* (1860).
Selections from the Poetical Remains of the late Sarah Martin (Yarmouth [1872]).
Selfe, Rose Emily. *Light amid London Shadows* (1906).
Sewell, Mrs [Margaret]. *'Thy Poor Brother': letters to a friend on helping the poor* (1863).
[Sheppard, Mrs Emma]. *Experiences of a Workhouse Visitor* (1857).
—— *An Out-Stretched Hand to the Fallen* [1860].
—— *Sunshine in the Workhouse* (1858).
Sims, George R. *et al. Sketches of the Salvation Army Social Work* (1906).
Sketch of the Origin and Results of Ladies' Prison Associations, with Hints for the Formation of Local Associations (1827).
[Smith, Theophilus]. *Statement of the Origin, Proceedings and Results of the Midnight Meetings for the Recovery of Fallen Women* [1860]
Speech of Mrs. Ormiston Chant at the Annual Meeting of the S. P. A., June 13th, 1883 [1883].
A Stallholder. *A Fancy Sale* [1875?].
Steer, Mary H. *Opals from Sand, a story of early days at the Bridge of Hope* [1912].
—— 'Rescue Work by Women among Women'. *Woman's Mission*, ed. Baroness Burdett-Coutts (1893), pp. 149–59.
Stephen, Sir James. *Essays in Ecclesiastical Biography*, 2 vols. (1849).
[Stevenson, Robert Louis]. *The Charity Bazaar: An Allegorical Dialogue* [1868].
Stodart, M. A. *Hints on Reading; addressed to a young Lady* (1839).
Sylvia's Book of Bazaars and Fancy Fairs [1879].
Talbot, James Beard. *The Miseries of Prostitution* (1844).

These Fifty Years. Being the Jubilee Volume of the London City Mission [1884].
Thomas, Edward V. *Twenty-Five Years' Labour among the Friendless and Fallen* (1879).
Thompson, Flora. *Lark Rise to Candleford* (1945).
Thornbury, G. W. and Walford, Edward. *Old and New London*, 6 vols. [1873 etc.].
Tod, Isabella M. S. *On the Education of Girls of the Middle Classes* (1874).
Trenholme, Edward C. *Rescue Work* (1927).
Tressell, Robert. *The Ragged Trousered Philanthropists* (1955).
Trimmer, Sarah. *The Œconomy of Charity*, 2 vols. (1801).
Twining, Louisa. 'The History of Workhouse Reform'. *Woman's Mission*, ed. Baroness Burdett-Coutts (1893), pp. 265–72.
—— 'On the Training and Supervision of Workhouse Girls'. *Transactions of the National Association for the Promotion of Social Science*, iii (1859), pp. 696–702.
—— *A Paper on the Condition of Workhouses* (1858).
—— *Readings for Visitors to Workhouses and Hospitals* [1865].
—— *Recollections of Workhouse Visiting and Management during twenty-five years* (1880).
—— *Workhouses and Pauperism* (1898).
—— *Workhouses and Women's Work* (1858).
—— 'Women as Official Inspectors'. *Nineteenth Century*, xxxv (1894), pp. 489–94.
—— 'Women as Public Servants'. *Nineteenth Century*, xxviii (1890), pp. 950–8.
Vanderkiske, R. W. *Notes and Narratives of a Six Years' Mission, principally among the dens of London* (1852).
A Visitor. *Mornings at the Union* (1858).
A Voice to the Young, from a Death Bed (1860).
Wakefield, Priscilla. *Reflections on the present condition of the female sex; with suggestions for its improvement* (1798).
Weigall, Mrs Arthur. *Seeking and Saving: being the rescue work of the Manchester Mission* ([Manchester] 1889).
Whateley, Miss E. J. *Cousin Mabel's Experiences: Sketches of Religious Life in England* [1870].
—— *Evangelical Teaching; its meaning and application* (1871).
[Wightman, Mrs J. B.]. *Annals of the Rescued* (1861).
Wilberforce, William. *A Practical View of the Prevailing Religious System of Professed Christians in the Higher and Middle Classes in this Country contrasted with real Christianity* (1797).
Wilkinson, James John Garth. *The Forcible Introspection of Women for the Army and Navy by the Oligarchy, considered Physically* (1870).
Williams, The Revd. Isaac. *Female Characters of Holy Scripture; in a series of sermons* (1859).

A Woman [Eliza Hinton]. *Thoughts for the Heart. Addressed to Women* (1847).
A Woman [Sarah Lewis]. *Woman's Mission* (1839).
A Woman. 'Woman's Rights and Duties'. *Blackwood's Magazine*, liv (1843), pp. 373-97.
A Woman. *Woman's Rights and Duties, considered with relation to their Influence on Society and on her own Condition*, 2 vols. (1840).
The Woman Question in Europe, ed. Theodore Stanton (1884).
Woman's Mission, ed. Baroness Burdett-Coutts (1893).
Woodrooffe, Anne. *Shades of Character: or, Mental and Moral Delineations; designed to promote the formation of the female character on the basis of Christian principle*, 2 vols. (1841).
The Works of the Rev. John Wesley, 14 vols. (1872).
A Younger Brother. *On Charity Bazaars* (1829).

VIII. Visiting Manuals etc.

Anon. *The District Visitor's Manual* (1840).
—— *District Visitor's Note Book and Parish Vade Mecum* (1866).
—— *District Visitors' Record* (1836).
—— *Hints for District Visitors* (1898).
—— *Hints on District Visiting* [1877].
—— *Hints to a Clergyman's Wife: or, Female parochial duties practically illustrated* (1832).
—— *The Ladies' Companion for Visiting the Poor* (1813).
—— *Useful Hints; gathered from the experience of a lifetime* [1880].
—— *A Woman's Secret; or, How to make Home Happy* (1860).
Barnett, Mrs S. A. *The Work of Lady Visitors* (n.d.).
Bosanquet, Charles B. P. *A Handy-Book for Visitors of the Poor in London* (1874).
A Clergyman's Daughter [Miss Charlesworth]. *The Female Visitor to the Poor* (1846).
D., W. A. *Hints, Helps, and Rules for District Visitors* (1895).
Davies, J. L. 'District Visiting'. *Lectures to Ladies on Practical Subjects* (Cambridge, 1855), pp. 117-34.
Hessey, Francis. *Hints to District Visitors* (1858).
A Lady. *The Cottager's Assistant; or, the Wedding Present* (1824).
Loch, C. S. *How to Help Cases of Distress* (1883).
Mitchell, Elizabeth Harcourt. *Lay Help in District Visiting* (1899).
Neil, The Revd. Charles. *The Christian Visitor's Handbook* (1882).
An Old Visitor. *Beneficent Visits in the Metropolis* (1817).
Sewell, Margaret A. *District Visiting* (1893).

IX. Contemporary Women's Memoirs, Autobiographies, Biographies etc.

Ada, Leila. *Leila Ada, the Jewish Convert. An Authentic Memoir*, by Osborne W. Trenery Heighway (1868).

Ainsworth, Sophia. *Mrs. Sophia Ainsworth (in religion Sister Mary Anne Ligouri of Jesus Crucified)*, by Anthony John Hanmer (1894).
Alford, Fanny. *Reminiscences; by a Clergyman's Wife*, ed. by the Dean of Canterbury [H. Alford] (1860).
Allen, Beatrice Jullian. *Our Sister Beatrice. Recollections of B. J. Allen and her Letters*, ed. by Grace Grier (1907).
Allen, Elizabeth. *The Faithful Servant; or, the History of Elizabeth Allen* (1824).
Allen, Hannah S. *A Beloved Mother. Life of Hannah S. Allen*, by her daughter (1884).
Allen, Maria. *Memoir of Mrs. Allen, of Woodhead-Hall, Staffordshire*, by her son [William S. Allen] (Manchester, 1872).
Alsop, Christine Majolier. *Memorials*, compiled by Martha Braithwaite (1881).
Anderson, Bessie. *The Morning of Life: A Memoir of Miss A--n, who was educated for a Nun*, by [Mary M. C. Methuen] (Bath [1851]).
Arbuthnot, Harriet. *The Journal of Mrs. Arbuthnot, 1820–1832*, 2 vols. (1950).
Ashford, Mary Ann. *Life of a Licensed Victualler's Daughter* (1844).
Ayres, Mary. *The Fruitful Bough. A Short Memoir of Mary Ayres*, by David Ives (Chalfont St. Peter [1854?]).
Baillie, Lady Grisell. *Recollections of a Happy Life* (Edinburgh, 1890).
Bairstow, Hannah. *Walking in the Light: A Memoir of Mrs. Hannah Bairstow, of Huddersfield, Yorkshire*, by the Revd. Thornley Smith (1868).
Balfour, Lady Frances. *Ne Obliviscaris Dinna forget*, 2 vols. [1930].
Ball, Frances. *Mrs. Ball, Foundress of the Institute of the Blessed Virgin Mary in Ireland and the British Colonies. A Biography*, by William Hutch (Dublin, 1879).
Ball, Hannah. *Memoir of Miss Hannah Ball*, ed. by John Parker (1839).
Barber, Miss Mary Ann Serrett. *Bread Winning; or, the Ledger and the Lute. An Autobiography* [1865].
Barclay, Florence L. *The Life of Florence L. Barclay. A Study in Personality*, by a daughter (1921).
Barratt, Edith. *Edith Barratt. Her Life and Thoughts* (1913).
Bartlett, Lavinia Strickland. *Mrs. Bartlett and her Class at the Metropolitan Tabernacle*, ed. by Edward H. Bartlett (1877).
Bass, Matilda. *She Walked with God, . . . Memorials of Mrs. Bass*, by [J. V. Bishop] (1881).
Bates, Mrs Ely. *Selections from the Correspondence of Mrs. Ely Bates, and incidents of her early Life*, by Mrs T. G. Tyndale, 2 vols. (Oxford, 1872).
Battersea, Constance. *Reminiscences* (1922).
Bayley, Elizabeth. *Piety and Usefulness a Hundred Years Old. A Memoir of Mrs. E. Bayley*, introduced by J. A. James (1856).
Bayly, Ada Ellen. *The Life of Edna Lyall (Ada Ellen Bayly)*, by J. M. Escreet (1904).

Beale, Dorothea. *Dorothea Beale of Cheltenham*, by Elizabeth Raikes (1908).
—— *Dorothea Beale*, by Elizabeth H. Shillito (London, 1920).
Beamish, Esther. *'A Voice that is Still'. Memorials of Esther Beamish*, by F. L. M. Beamish [1885].
Bennett Sarah. *The Christian Governess: A Memoir and Selection from the Correspondence of Miss Sarah Bennett*, by George B. Bennett (1862).
Besant, Annie. *An Autobiography* (1893).
Blake, Mrs Alice Elizabeth. *Memoirs of a Vanished Generation, 1813–1855* (1909).
Blake, Dame Louisa Aldrich. *Dame Louisa Aldrich-Blake*, by Lord Riddell [1926].
Booker, Beryl Lee. *Yesterday's Child, 1890–1909* (1937).
Booth, Catherine. *The Life of Catherine Booth, the Mother of the Salvation Army*, by Frederick St. George de Lautour Booth Tucker, 2 vols. [1893].
Bray, Anna Eliza. *Autobiography of Anna Eliza Bray*, ed. by John A. Kempe (1884).
Britten, Emma Hardinge. *Autobiography of Emma Hardinge Britten*, ed. by Mrs Margaret Wilkinson (1900).
Broad, Lucy. *A Woman's Wanderings the World over* [1909].
Broadley, Charlotte. *Mother Charlotte (Mrs. Broadley of Carnmenellis). A Sketch*, by Louisa Herbert [1882].
Bromiley, Susanna. *The Gathered Flower: A Memoir of Miss Susanna Bromiley*, by the Revd. T. W. Aveling (1857).
Broomfield, Hannah. *A Golden Sunset: Being an Account of the last Days of Hannah Broomfield* (1874).
Brown, Mary Crawford. *Mary Crawford Brown. A Memoir*, by James Strahan (1920).
Bryan, Ruth. *'Handfuls of Purpose'; or, Gleanings from the Inner Life of R. Bryan* (1862).
—— *Letters of Ruth Bryan* (1865).
Budgett, Sarah. *A Memoir of the late Mrs. Sarah Budgett*, by John Gaskin (1840).
Bulmer, Agnes. *Memoir of Mrs. Agnes Bulmer*, by Anne Ross Collinson (1837).
—— *Select Letters of Mrs. Agnes Bulmer*, introduced by the Revd. William M. Bunting (1842).
Bundy, Mrs. *Short Memorials of Widow Bundy, an aged Christian in Humble Life*, by E. Jenyns (Brighton, 1853).
Burton, Agnes. *Personal Service. Being a Short Memoir of Agnes Burton*, by Ellen Maples (1913).
Burton, Margaret. *Life and Correspondence of the Late Mrs. Margaret Burton*, by John Dungett (Darlington, 1832).
Buss, Frances Mary. *Frances Mary Buss and her Work for Education*, by Annie E. Ridley (1895).

Butler, Josephine E. *Josephine E. Butler, an Autobiographical Memoir*, eds. George W. and Lucy A. Johnson (Bristol and London, 1909).

Buxton, Lady Victoria. *Lady Victoria Buxton*, by G. W. E. Russell (1919).

C., Mary Elizabeth. *'Leaning on her Beloved'. A Memorial of M. E. C.* (1857).

Cameron, Lucy Lyttelton. *The Life of Mrs. Cameron: Partly an Autobiography, and from her Private Journals*, ed. by her son [1862].

Campbell, Isabella. *Peace in Believing: A Memoir of Isabella Campbell*, by Robert Story (1854).

Campbell, Lady Victoria. *Lady Victoria Campbell, a Memoir*, by Lady Frances Balfour [1911].

Camplin, Sarah. *Memorial of Mrs. Camplin*, ed. by Eliza T. Tooth (1833).

Cappe, Catharine. *Memoirs of the Life of the late Mrs. Catharine Cappe* (1822).

Carbery, Mary. *Happy World. The Story of a Victorian Childhood* (1941).

Cargill, Margaret. *Memoirs of Mrs. Margaret Cargill*, by David Cargill (1841).

Carpenter, Mary. *The Life and Work of Mary Carpenter*, by J. E. Carpenter (1879).

Carr, Mrs Richard. *Memoirs and Correspondence of Mrs. Richard Carr, of Birmingham*, by the Revd. Jacob Stanley (1845).

Cavendish, Lady Frederick. *The Diary of Lady Frederick Cavendish*, ed. by John Bailey, 2 vols. (1927).

Chadwick, Ellen. *Ellen Chadwick, the Famous Manchester Invalid*, by H. B. Harrison (Manchester, 1912).

Charlotte, Princess. *A Brief Memoir of the Princess Charlotte of Wales*, by Lady Rose Weigall (1874).

Chatelain, Marie. *Sister Chatelain or Forty Years' Work in Westminster*, ed. by Lady Amabel Kerr (1900).

Chatterton, Lady (Georgiana). *Memoirs of Georgiana, Lady Chatterton. With some Passages from Her Diary*, by Edward Heneage Dering (1878).

Clarke, Anna Maria. *A Memoir of Anna Maria Clarke*, by the Revd. Thomas Grey Clarke (1853).

Clarke, Mrs Mary Cowden. *My Long Life: An Autobiographical Sketch* (1896).

Clayton, Mrs George. *The Life of Mrs. George Clayton*, by Joseph Sortain (1844).

Clibborn, Catherine Booth. *The Maréchale*, by James Strahan (1913).

Clifford, Mary. *Mary Clifford*, by Gwen Mary Williams (Bristol, 1920).

Clive, Caroline. *Caroline Clive. From the Diary and Family Papers of Mrs. Archer Clive (1801–1873)*, ed. by Mary Clive (1949).

Clough, Anne Jemima. *A Memoir of Anne Jemima Clough*, by Blanche Athena Clough (1897).

Clough, Margaret M. *Extracts from the Journal and Correspondence of the late Mrs. M. M. Clough* (1829).
Cobb, Mary. *Extracts from the Diary and Letters of Mrs. Mary Cobb* (1805).
Cobbe, Frances Power. *Life of Frances Power Cobbe*, 2 vols. (1894).
Connolly, Jane. *Old Days and Ways* (1912).
Cooper, Mary. *Memoirs of the late Mrs. Mary Cooper, of London* (1814).
Cooper, Mary Ann. *The Root and the Branches. Memoirs of Mrs. M. A. Cooper and her two Grandchildren, Emma and Sarah Ann Cooper*, by John Cooper (1854).
Cooper, Mary Anna. *Memorials of a Beloved Mother: Being a Sketch of the Life of Mrs. Cooper*, by [Charlotte Bickersteth Wheeler] (1853).
Cooper, Mary Sarson. *Memorials of Mrs. Mary Sarson Cooper . . . compiled from her Diary and Correspondence*, by Henry Fish (1855).
Cooter, Eliza. *Light in Darkness: A Short Account of a Blind Deaf Mute*, by Sarah Robinson (1859).
—— *The Darkness Past*, by [Sarah Robinson] (1861).
Coutts, Baroness Burdett. *A Sketch of her Public Life and Work* (Chicago, 1893).
Cox, Adelaide. *Hotchpotch* (1937).
Croad, Rebecca Caroline Hayman. *A Service of Suffering: or, Leaves from the Biography of Mrs. Croad*, by J. G. Westlake [1885?].
Crosby, Fanny J. *Memories of Eighty Years* (1907).
Cryer, Mary. *The Devotional Remains of Mrs. Cryer* (1854).
—— *Holy Living; Exemplified in the Life of Mrs. Mary Cryer*, by the Revd. Alfred Barrett (1845).
Cumming, Miss Constance Frederica Gordon. *Memories* (Edinburgh, 1904).
Cusack, Margaret Anna. *Five Years in a Protestant Sisterhood and Ten Years in a Catholic Convent* (1869).
Cusack, Mary Frances. *The Story of my Life* (1891).
Daniell, Georgiana Fanny Shipley. *A Soldier's Daughter. A Short Memorial of Miss Daniell of Aldershot* (1894).
Daniell, Louisa. *Aldershot: A Record of Mrs. Daniell's Work amongst Soldiers, and its Sequel*, by G. F. S. Daniell (1879).
Davidson, Elizabeth. *Memoir of Miss Elizabeth Davidson*, by the Revd. John Clunie (1813).
Davies, Mrs William. *Brief Memorials of the late Mrs. William Davies, of Canterbury*, by William Davies (1842).
Davis, Ann. *Memorials of Ann Davis, an Eminent Christian in Humble Life*, by a Visitor (1849).
Davis, Elizabeth. *The Autobiography of Elizabeth Davis, a Balaclava Nurse*, ed. by Jane Williams, 2 vols. (1857).
Dawson, Sarah, *A Brief Memoir of Mrs. Sarah Dawson*, by Alfred Dawson [1828].

Dening, Mrs Henry. *'She Spake of Him', being Recollections of the loving labours and early death of the late Mrs. Henry Dening*, by Mrs Grattan Guinness (Bristol [1872]).

Doudney, Dinah. *A Child's Memorial; or, a new token for children containing an account of the early piety and happy death of Miss Dinah Doudney*, by the Revd. John Griffin (1856).

Drummond, Maria. *Maria Drummond: A Sketch*, by C. Kegan Paul (1891).

Dyer, Mary. *Christhood: As seen in the Life Work of Mary Dyer*, by the Revd. George Dyer [1889].

Duncan, Mary Lundie. *Memoir of Mary Lundie Duncan: Being Recollections of a Daughter by her Mother*, by [Mrs Lundie] (Edinburgh, 1854).

Duncan, Mrs W. W. *Memoir of Mrs. W. W. Duncan; being Recollections of a Daugher*, by her mother [Mrs Lundie] (Edinburgh, 1841).

Dyson, Mrs Charles. *Memorials of a departed Friend*, ed. by [Charles Dyson] (1833).

Earle, Mrs C. W. *Memoirs and Memories* (1911).

East, Ann. *Peace in Believing, exemplified in a Memoir of Ann, the beloved Wife of the Rev. John East* (Bristol, 1837).

Eastlake, Lady (Elizabeth Rigby). *Journals and Correspondence of Lady Eastlake*, ed. by Charles Eastlake Smith, 2 vols. (1895).

Edward, Catherine. *Missionary Life among the Jews in Moldavia, Galicia and Silesia. Memoir and Letters of Mrs. Edward* (1867).

Edward, Eliza. *Diary of a Quiet Life* (1887).

Elliott, Charlotte. *Leaves from the unpublished Journals, Letters and Poems of Charlotte Elliott* [1874].

Ellis, Sarah Stickney. *The Home Life and Letters of Mrs. Ellis*, compiled by her nieces [1893].

Farjeon, Eleanor. *A Nursery in the Nineties*, (1935).

Farningham, Marianne. *A Working Woman's Life. An Autobiography* (1907).

Fawcett, Millicent Garrett. *What I Remember* (1924).

Ffoulkes, Maude M. C. *My Own Past* (1915).

Fletcher, Eliza. *A Woman's Work: Memorials of Eliza Fletcher*, by the Revd. Charles Adamson Salmond (1885).

Fletcher, Mary. *The Life of Mrs. Fletcher*, by Thomas E. Gill 1845).

—— *The Life of Mrs. Mary Fletcher*, ed. by Henry Moore (1818).

Forbes, Lady Angela. *Memories and Base Details* [1922].

Fox, Eliza. *Memoir of Mrs. Eliza Fox*, ed. by Franklin Fox (1869).

Fox, Maria S. *Memoirs of Maria Fox, late of Tottenham, consisting chiefly of extracts from her Journals and Correspondence* (1846).

Freeman, Ann. *A Memoir of the Life and Ministry of Ann Freeman* (Exeter, 1831).

Freeman, Clara. *A Brief Memorial of Clara Freeman, to which is added some of her Bible Studies*, by Annie M. Freeman [1913].

Fullerton, Lady Georgiana. *The Inner Life of Lady Georgiana Fullerton*, by [Fanny M. Taylor] [1899].

Fullerton, Lady Georgiana. *Life of Lady Georgiana Fullerton*, by Pauline Craven, translated from French by Henry James Coleridge (1888).

Furse, Dame Katharine. *Hearts and Pomegranates. The Story of Forty-Five Years, 1875–1920* (1940).

Garfit, Lisa. *In Memoriam Lisa Garfit*, by Alice M. Bethune (1887).

Gaskell, Elizabeth. *The Letters of Mrs. Gaskell*, eds. J. A. V. Chapple and Arthur Pollard (Manchester, 1966).

Gibbs, Via. *Via Gibbs. A Memoir*, by Madeline Alston (1921).

Gibson, Jane. *Memoirs of Mrs. Jane Gibson*, by Francis A. West (1837).

Gilbert, Ann. *Autobiography and other Memorials of Mrs. Gilbert*, ed. by Josiah Gilbert, 2 vols. (1874).

Gilbert, Elizabeth. *Elizabeth Gilbert and Her Work for the Blind*, by Frances Martin (1887).

Gilpin, Mary Ann. *Memoir of Mary Ann Gilpin* (1841).

Gladstone, Catherine. *Catherine Gladstone*, by Mary Drew (1919).

Gladstone, Mary. *Mary Gladstone (Mrs. Drew). Her Diaries and Letters*, ed. by Lucy Masterman (1930).

Goodman, Margaret. *Experiences of an English Sister of Mercy* (1862).

Graham, Mary Jane. *A Memoir of Miss Mary Jane Graham*, by the Revd. Charles Bridges (1832).

Grane, Miss. *Memoirs of Miss Grane, illustrative of the nature and effects of Christian principles, compiled principally from her own papers* (1842).

Green, Harriet. *Harriet Green*, by Sophia M. Fry (1903).

Greenwell, Dora. *Memoirs of Dora Greenwell*, by William Dorling [1885].

Grey, Mary. *Brief Sketches of the early history, conversion, and closing period of the life of Mary, second daughter of the Hon. John Grey*, by [Henry C. Grey], [1855].

Gundry, Maria. *Extracts from the Letters and Memoranda of Maria Gundry* (1847).

Gurney, Priscilla. *Memoir of Priscilla Gurney*, ed. by Susanna Corder (1856).

Haldane, Mary Elizabeth. *Mary Elizabeth Haldane. A Record of a Hundred Years (1825–1925)*, ed. by Elizabeth Sanderson Haldane [1925].

Hamilton, Elizabeth. *Memoirs of the late Mrs. Elizabeth Hamilton*, by Elizabeth Ogilvie Benger, 2 vols. (1818).

Hanbury, Charlotte. *Autobiography*, ed. by Mrs Albert Head (1901).

Harris, Amy and Rosalie. *Amy and Rosalie. A Mother's Memorials of Two Beloved Children*, by [Emily Marion Harris] (1854).

Harrison, Ann. *Memoranda of the late Ann Harrison, of Weston*, ed. by the Revd. T. Best (Sheffield, 1859).

Havergal, Frances Ridley. *Memorials of Frances Ridley Havergal*, by M. V. G. H[avergal] (1882).

Havergal, Miss Maria Vernon Graham. *The Autobiography of M. V. G.*

Havergal, with Journals and Letters, ed. by J. M. Crane (Edinburgh, 1887).

Head, Mrs A. *Life of Mrs. Albert Head*, by Charlotte Hanbury [1905].

Henwood, Loveday. *Extracts from the Memoir and Letters of the late Loveday Henwood* (1847).

Herschell, Helen S. *'Far above Rubies'. Memoir of Helen S. Herschell*, by her daughter, ed. by Ridley H. Herschell (1854).

Higginson, Teresa Helena. *Teresa Helena Higginson. Servant of God. 'The Spouse of the Crucified' 1844–1905*, by Lady Anne Cecil Kerr (1927).

Hill, Octavia. *Life of Octavia Hill as told in her letters*, ed. by C. Edmund Maurice (1913).

Hill, Rosamond Davenport. *Memoir of Rosamond Davenport-Hill*, by Ethel E. Metcalfe (1904).

Hilton, Marie. *Marie Hilton, Her Life and Work 1821–1896*, by J. Deane Hilton (1897).

Hinderer, Anna. *Seventeen Years in the Yoruba Country* [1877].

Hirst, Alice Ellen. *The Recollections of a Minister's Wife* [1904].

Hoare, Rachel. *Christian Experience: a Memoir of Mrs. Hoare*, ed. by W. Worth Hoare (1852).

Hoare, Sophia F. *Early Religion; or, a Memoir of Sophia F. Hoare* (Birmingham [1855?]).

Hobhouse, Margaret. *Margaret Hobhouse and her Family*, by Stephen Hobhouse (Rochester, 1934).

Hobson, Catharine Leslie. *Catharine Leslie Hobson, Lady-Nurse, Crimean War, and her Life*, by the Revd. William Fraser Hobson (1888).

Hodgkin, Elizabeth. *Extracts from Familiar Letters of the late Elizabeth Hodgkin* (1842).

Holmes, Elizabeth. *A Sister's Record; or, Memoir of Mrs. Marcus H. Holmes* (1844).

Holy, Lucy Maria. *Memorials of Lucy Maria Holy* (1872).

Hopkins, Mrs Eliza Ann. *Some account of the Religious Experience, last Illness, and Death of E. A. Hopkins* (Brigg, 1866).

Hopkins, Ellice. *Ellice Hopkins: a memoir*, by Rosa M. Barrett (1907).

—— *An English Woman's Work among Workingmen* (1875).

Houghton, Sarah. *Family Memorials of the late Mr. & Mrs. R. Houghton, of Huddersfield; and of several of their Children*, by [Miss H. Houghton] (1846).

Howard, Lady Etheldreda Fitzalan. *Memoir of a Sister of Charity—Lady Etheldreda Fitzalan Howard*, by Lady Anne Cecil Kerr (1928).

Howitt, Mary. *An Autobiography*, ed. by Margaret Howitt (1880).

Hubbard, Louisa M. *A Woman's Work for Women: Being the aims, efforts and aspirations of Louisa M. Hubbard*, by Edwin A. Pratt (1898).

Hughes, M. Vivian. *A London Child of the Seventies* (1934).

—— *A London Girl of the Eighties* (1936).

Hunt, Agnes. *Reminiscences* (Shrewsbury, 1935).

Hunt, Ann. *Memorials and Letters of Ann Hunt*, ed. by Matilda Sturge (1898).
Innes, Martha. *Memoir of Mrs. Martha Innes: with extracts from her Diary and Letters*, ed. by William Innes (1844).
Jameson, Anna. *Memoirs of the Life of Anna Jameson*, by Gerardine Macpherson (1878).
Jarrett, Rebecca. *Rebecca Jarrett*, by Josephine Butler [1885].
Jenour, Ann. *The Christian Wife and Mother, being a Brief Memoir of Ann Jenour*, by the Revd. Alfred Jenour (1840).
Jerram, Hannah. *A Tribute of Parental Affection to the Memory of a beloved and only Daughter*, by Charles Jerram (1824).
Jeune, Margaret Dyne. *Pages from the Diary of an Oxford Lady, 1843-1862*, ed. by Margaret Jeune Gifford (Oxford, 1932).
Jeune, Susan Elizabeth Mary (Baroness St. Helier). *Memories of Fifty Years* (1909).
Johnston, Blanche Read. *The Lady with the Other Lamp. The Story of Blanche Read Johnston*, by Mary Morgan Dean (Toronto [1919]).
Johnstone, Ann. *Memoir of the late Mrs. Ann Johnstone* (Edinburgh, 1846).
Jones, Agnes Elizabeth. *Memorials of Agnes Elizabeth Jones*, by [J. Jones] (1871).
Judson, Ann H. *A Sketch of Mrs. Ann H. Judson*, by Clara Lucas Balfour (1854).
Jukes, Harriet Maria. *The Earnest Christian. Memoir, Letters, and Journals of Harriet Maria, Wife of the late Rev. Mark R. Jukes*, ed. by Mrs H. A. Gilbert (1858).
K., M. *Memorials of a departed Friend; consisting of selections from her letters* (1867).
Kell, Mrs Edmund. *Memorials of the Rev. Edmund Kell, B. A., F. S. A., and Mrs. Kell, of Southampton* (1875).
Kenning, Elizabeth. *Some account of the Life of Elizabeth Kenning, chiefly drawn up by herself* (Bradford [1829]).
Kerr, Henrietta. *The Life of Mother Henrietta Kerr, Religious of the Sacred Heart*, ed. by John Morris (1887).
Kilham, Hannah. *Extracts from the Letters of Hannah Kilham* (1831).
— *Memoirs of the late Hannah Kilham; chiefly compiled from her Journal*, ed. by Sarah Biller (1837).
— *Sketch of the Life of Hannah Kilham*, by Hannah Maria Wigham [1880].
King, Mrs Elizabeth. *Lord Kelvin's Early Home* (1909).
King, Elizabeth Thomson King. *My Sister*, by Agnes Gardner King (Glasgow, 1914).
Kinnaird, Emily. *Reminiscences* (1925).
Kinnaird, Mary Jane. *Mary Jane Kinnaird*, by Donald Fraser (1890).
Knapp, Susanna. *Fruits of Righteousness in the Life of Susanna Knapp*, by Edith Rowley (1866).
Lambert, Hetty. *Strength Perfected in Weakness: being Memorials of Hetty Lambert*, by William John Townsend [1893].

Leece, Sophia. *Narrative of the Life of Miss Sophia Leece*, by the Revd. Hugh Stowell (Liverpool, 1820).

Legge, Mary. *A Life of Consecration. Memorials of Mrs. Mary Legge*, by a son (1883).

Lewis, Georgina King. *An Autobiographical Sketch*, ed. by Barbara Duncan Harris (1925).

Linton, Lynn. *Mrs. Lynn Linton. Her Life, Letters, and Opinions*, by George Somes Layard (1901).

Lloyd, Anna. *Anna Lloyd, 1837–1925. A Memoir*, ed. by Edyth M. Lloyd (1928).

Lucas, Margaret Bright. *Margaret Bright Lucas. The Life Story of a 'British Woman'*, by Henry J. B. Heath (1890).

Lytton, Lady (Rosina). *Life of Rosina, Lady Lytton*, by Louisa Devey (1887).

McDougall, Harriette. *An Early Victorian Heroine*, by Mary Bramston (1911).

Mackenzie, Anne. *An Elder Sister. A Short Sketch of Anne Mackenzie, and her Brother the Missionary Bishop*, by Frances Awdry (1878).

McMillan, Rachel. *The Life of Rachel McMillan*, by Margaret McMillan (1927).

McNeill, Elizabeth Wilson. *Memoir of the Right Hon. Sir John McNeill, G. C. B., and of his second wife Elizabeth Wilson*, by [Florence MacAlister] (1910).

Macpherson, Annie. *God's Answers; A Record of Miss Annie Macpherson's Work at the Home of Industry, Spitalfields, London, and in Canada*, by Clara M. S. Lowe (1882).

Mactaggart, Mrs Ann. *Memoirs of a Gentlewoman of the Old School*, by a Lady, 2 vols. (1830).

Malleson, Elizabeth. *Elizabeth Malleson, 1828–1916. Autobiographical Notes and Letters. With a Memoir by Hope Malleson* ([London] 1926).

Mann, Fannie Glover. *Memorials of Fannie Glover Mann* (Sheffield, 1886).

Marsh, Catherine. *The Life and Friendships of Catherine Marsh*, by L. E. O'Rorke (1917).

Martin, Sarah. *A Brief Sketch of the Life of the late Miss Sarah Martin* (Yarmouth, 1845).

—— *Sarah Martin,the Prison-Visitor of Great Yarmouth*, by [George Mogridge] ([1872]).

Martindale, Hilda, *et al. From One Generation to Another, 1839–1944. A Book of Memoirs* (1944).

Maurice, Anne and Emma. *Memorials of Two Sisters* (1833).

Meath, Mary Countess of. *The Diaries of Mary Countess of Meath*, ed. by Reginald Brabazon, 2 vols. [1928].

Meredith, Susanna. *Saved Rahab! An Autobiography* (1881).

—— *Susanna Meredith. A Record of a Vigorous Life*, by Mary Anne Lloyd (1903).

Meuricoffre, Harriet. *In Memoriam Harriet Meuricoffre*, by Josephine Butler [1901].

Mitchell, Hannah. *The Hard Way Up: The Autobiography of Hannah Mitchell*, ed. by Geoffrey Mitchell (1977).

Monsell, Harriet. *Harriet Monsell, a Memoir*, by T. T. Carter (1884).

Montefiore, Dora B. *From a Victorian to a Modern* (1927).

More, Hannah. *Memoirs of the Life and Correspondence of Mrs. Hannah More*, by William Roberts, 4 vols. (1834).

Mortimer, Elizabeth. *Memoirs of Mrs. Elizabeth Mortimer*, by Agnes Bulmer (1836).

Mundy, Louisa. *A Brief Memoir of the late Mrs. Mundy, . . . Missionary of the London Missionary Society*, by the Revd. G. Mundy (Calcutta, 1843).

Nevill, Lady Dorothy. *My Own Times*, ed. by Ralph Nevill (1912).

—— *The Reminiscences of Lady Dorothy Nevill*, ed. by Ralph Nevill (1906).

Newell, Harriet. *Memoirs of Mrs. Harriet Newell*, by Leonard Woods (1818).

Newton, Adelaide Leaper. *A Memoir of Adelaide Leaper Newton*, by the Revd. John Baillie (1856).

Newton, Elizabeth. *Memorials of the Life of Mrs. Newton*, by her daughter (1867).

Nicholson, Rosa E. C. *A Brief Memoir of the late Miss Rosa E. C. Nicholson*, by the Revd. W. H. Krause (Dublin, 1853).

Oliphant, Margaret. *The Autobiography and Letters of Mrs. M. O. W. Oliphant*, ed. by Mrs Harry Coghill (Edinburgh and London, 1899).

Opie, Amelia. *Memorials of the Life of Amelia Opie*, by Cecilia Lucy Brightwell (Norwich, 1854).

Orford, Hannah. *Hannah Orford* [1854].

Ormerod, Eleanor. *Eleanor Ormerod, LL.D.*, ed. by Robert Wallace (1904).

Palmer, Louisa. *Christian Devotedness: or, Memorials of Mrs. and Miss Palmer, of Newbury*, by Henry March (1844).

Panton, Jane Ellen. *Leaves from a Life* (1908).

—— *More Leaves from a Life* (1911).

Parris, Mary and Hephzibah. *Memoirs of Mary and Hephzibah Parris, and a Brief Memoir of Miriam Parris*, by a sister (1858).

Parsloe, Muriel Jardine. *A Parson's Daughter* (1935).

Pattison, Dora. *Sister Dora. A Biography*, by Margaret Lonsdale (1880).

Pearson, Jane. *Memoir of the Life and Religious Experiences of Jane Pearson* (York, 1839).

Pease, Mary. *Mary Pease. A Memoir*, by Marion E. Fox (1911).

Peck, Winifred. *A Little Learning; or A Victorian Childhood* (1952).

Peel, Lady Georgiana. *Recollections of Lady Georgiana Peel*, compiled by Ethel Peel (1920).

Petrie, Irene. *Irene Petrie. Missionary to Kashmir*, by Mrs Ashley Carus Wilson (1900).

Phipson, Rosalinda. *Youthful Consecration; a Memorial of Rosalinda Phipson*, introduced by J. A. James (1844).
Polhill, Mrs Eleanor Agnes. *Willing and Obedient: Memorials of E. A. Polhill*, by Annie W. Marston [1905].
— *With the King. Pages from the Life of Mrs. Cecil Polhill*, by Annie W. Marston [1905].
Poulson, Sarah. *My Mother's Memoir: A Record of the Providence of God*, by Edward Poulson (1885).
Poulson, Sophy. *The Life of Sophy Poulson: A true love story of a good and beautiful girl*, by Edward Poulson [1887].
Priestley, Lady (Eliza). *The Story of a Lifetime* (London, 1904).
Prust, Mrs Edmund Thornton. *Memorials of the Life of Mrs. Edmund T. Prust, of Northampton*, by Edmund T. Prust [1866?].
Pryor, Mary. *Mary Pryor. A Life Story of a Hundred Years Ago*, by Mary Pryor Hack (1887).
Radcliffe, Mary Ann. *The Memoirs of Mrs. M. A. Radcliffe in familiar Letters to her Female Friend* (Edinburgh, 1810).
Ramsay, Martha Laurens. *Memoirs of the Life of Martha Laurens Ramsay*, by David Ramsay (Glasgow, 1818).
Renton, A. *A Brief Memorial of the late Mrs. Renton*, by [William Renton] (1852).
Ripley, Dorothy. *The Extraordinary Conversion and Religious Experience, of Dorothy Ripley* (1817).
Robertson, Henrietta. *Memorials of Henrietta Robertson*, ed. by Anne Mackenzie (Cambridge, 1866).
Robinson, Sarah. *Active Service; or, Work among our Soldiers*, by Ellice Hopkins (1872).
— *The Soldier's Friend. A Pioneer's Record* (1913).
Robson, Mrs J. T. *Sunset at Noonday. Memorials of Mrs. J. T. Robson, of Hull*, by the Revd. Joseph Wood (1871).
Rogers, Hester Ann. *The Experience and Spiritual Letters of Mrs. Hester Ann Rogers* (1857).
Ross, Eliza Scott. *Letters to Children. A Narrative of the Life and Death of Eliza Scott Ross*, by Thomas Mann (1847).
Ross, Janet. *Early Days Recalled* (1891).
Rothschild, Lady Constance de. *Lady de Rothschild and her Daughters, 1821–1931*, by Lucy Cohen (1935).
Row, Susanna. *Memoir of the late Miss Susanna Row, of Cardington, Bedfordshire*, by Sarah Smith Jones (Hexham [1867]).
Sandes, Elise. *Enlisted; or My Story* (Cork and London, 1896).
Savage, Sarah. *Memoirs of the Life and Character of Mrs. Sarah Savage* (1819).
Scott, Mary. *Memoir of Mary Scott*, by the Revd. Thomas Scott [1855?].
Sellon, Priscilla Lydia. *Miss Sellon and the Sisters of Mercy*, by Diana A. G. Campbell (1852).
Sewell, Elizabeth M. *The Autobiography of Elizabeth M. Sewell*, ed. by Eleanor L. Sewell (1907).

Sewell, Mary. *The Life and Letters of Mrs. Sewell*, by Mrs Mary Bayly (1889).

Sharmon, Charlotte. *Charlotte Sharmon: The Romance of a Great Faith*, by Marguerite Williams [1931].

Shenston, Mary. *Memoir and Select Remains of Miss Mary Shenston*, by her Brother and Sister (1823).

Shepherd, Mary. *Miss Shepherd, of Cheadle, Staffordshire*, by the Revd. Buckley Yates (Manchester, 1876).

Sherman, Martha. *The Pastor's Wife. A Memoir of Mrs. Sherman*, by James Sherman (1848).

Sherrington, Mary Ignatius. *Mother Mary Ignatius Sherrington*, by John William Gilbert (1915).

Shore, Emily. *Journal of Emily Shore* (1891).

Smith, Jane. *The Trial of Faith; or, a Brief Memoir of Jane Smith*, by the Revd. James Jerram (1849).

Smith, Mary. *The Autobiography of Mary Smith, Schoolmistress and Nonconformist* (1892).

Smith, Mrs Reginald. *A Memorial Sketch of the Life and Labours of Mrs. Reginald Smith* (Dorchester, 1877).

Smith, Mrs Thornley. *A Christian Mother: Memoirs of Mrs. Thornley Smith*, by [Thornley Smith] (1885).

Somerset, Lady Henry. *Lady Henry Somerset*, by Kathleen Fitzpatrick (1923).

Southall, Eliza. *Portions of the Diary, Letters and other Remains of Eliza Southall* (Birmingham, 1855).

Spurgeon, Mrs C. H. *Ten Years of My Life in the Service of the Book Fund* (1887).

Squirrell, Elizabeth. *The Autobiography of Elizabeth Squirrell* (1853).

Stanley, Catherine. *Memoirs of Edward and Catherine Stanley*, ed. by Arthur Penrhyn Stanley (1880).

Stanley, Maria and Henrietta. *The Ladies of Alderley. Being the Letters between Maria Josepha, Lady Stanley of Alderley and her Daughter-in-law Henrietta Maria Stanley during the Years 1841–1850*, ed. by Nancy Mitford (1938).

Stannard, Mrs M. *Memoirs of a Professional Lady Nurse* (1873).

Stanton, Lucy Ann. *A Short Account of the Life of Mrs. Lucy Ann Stanton*, by her son Vincent Henry Stanton (Halseworth, 1883).

Statham, Louisa Maria. *Memoir of Louisa Maria Statham*, by the Revd. John Statham (1842).

Stevenson, Louisa and Flora. *Recollections of the Public Work and Home Life of Louisa and Flora Stevenson* (Edinburgh [1914]).

Stewart, Louisa Hooper. *Louisa. Memories of a Quaker Childhood*, ed. by Evelyn Roberts (1970).

Stone, Matilda L. A. *In Memory of our Dear Friend, Miss Stone*, by C. H. Dight ([1885?]).

Story, Janet Leith. *Early Reminiscences* (Glasgow, 1911).

Strickland, Agnes. *Life of Agnes Strickland*, by Jane Margaret Strickland (1887).

Sturge, Elizabeth. *Reminiscences of my Life* (Bristol, 1928).

Sumner, Mary. *Mary Sumner, her Life and Work*, by Mrs [Mary] Porter (Winchester, 1921).

Sutton, Charlotte. *Memoir of Mrs. Sutton; late Missionary to Orissa, East Indies* [1830?].

Swallow, Jane. *An Old Methodist; or, Memoirs of Mrs. Jane Swallow*, by Thomas Swallow (Liverpool [1861]).

T., M. M. *Her Record is on High. A Simple Memorial of M. M. T.* (1855).

Tait, Catharine. *Catharine and Craufurd Tait*, ed. by the Revd. William Benham (1879).

Tatton, Elizabeth. *The Officer's Daughter: A Memoir of Miss Elizabeth Tatton*, by the Revd. Octavius Winslow (Edinburgh, 1848).

Taylor, Ann. *Spiritual Life delineated and exemplified: A Memorial of Miss Ann Taylor, of Halifax*, by George Turner (1851).

Taylor, Fanny Margaret. *Mother Mary Magdalen of the Sacred Heart (Fanny Margaret Taylor) Foundress of the Poor Servants of the Mother of God, 1832–1900*, by Francis Charles Devas (1927).

Taylor, Jane. *Jane Taylor: Her Life and Letters*, by Mrs H. C. Knight (1880).

Taylor, Rebekah H. *Letters of Mrs. H. W. Taylor to Members of her Classes, and Friends*, ed. by Herbert W. Taylor [1878].

Thomas, Mrs N. *Christ Magnified: The Life of Mrs. N. Thomas*, by the Revd. David Davies [1884].

Timms, Mary. *Memoirs of the late Mrs. Mary Timms*, ed. by the Revd. E. Morgan (1835).

Tonna, Charlotte Elizabeth. *Memoir of Charlotte Elizabeth* (Bristol [1852]).

Townley, Charlotte. *A Brief Memoir of Charlotte Townley*, by C. G. Townley (Limerick, 1839).

Tremenheere, Camilla Eliza. *Camilla Eliza Tremenheere. A Memoir*, by Seymour Grieg Tremenheere (1922).

Trimmer, Sarah. *Some Account of the Life and Writings of Mrs. Trimmer with Original Letters, and Meditations and Prayers*, 2 vols. (1814).

Tucker, Charlotte Maria. *A Lady of England. The Life and Letters of Charlotte Maria Tucker*, by Agnes Giberne (1895).

Tuckett, Sarah Ann. *The Youthful Disciple. A Memoir of Sarah Ann Tuckett*, by the Revd. Thomas Haynes (1845).

Tweddell, Eliza. *The Saviour's Grace and Truth in the Life of Mrs. Tweddell* (Liverpool, 1830).

Twining, Louisa. *Recollections of Life and Work* (1893).

—— *Supplement to 'some facts in the history of the Twining family'*, (Salisbury, 1893).

Urmston, Harriett E. H. *The Starry Crown, a Sketch of the Life Work of Harriett Urmston*, by George Everard (1898).

Vachell, Ada. *Ada Vachell of Bristol*, by F. M. Unwin (Bristol, 1928).

Wake, Lady (Charlotte). *The Reminiscences of Charlotte, Lady Wake*, ed. by Lucy Wake (1909).

Wakefield, Rebecca. *Memoirs of Mrs. Rebecca Wakefield, Missionary in East Africa*, by Robert Brewin (1888).

Walsh, Anna Maria Drummond. *Dear Annie. A Brief Memorial of A. M. D. Walsh* [1855].

Ward, Mrs. *My Mother; or, Home scenes in Yorkshire*, by Annie Ward (1866).

Ward, E. M. *Mrs. E. M. Ward's Reminiscences*, ed. by Elliott O'Donnell (1911).

Ward, Mrs Humphrey. *Mrs. Humphrey Ward: Her Work and Influence*, by J. Stuart Walters (1912).

Waring, Mary. *A Diary of the Religious Experience of Mary Waring* (1809).

Watkins, Ann. *Extracts from the Memoranda and Letters of Ann Watkins, a Minister of the Society of Friends* (Ipswich, 1888).

Webb, Beatrice. *My Apprenticeship* (1926).

Weeton, Ellen. *Miss Weeton. Journal of a Governess, 1807–1811*, ed. by Edward Hall, 2 vols. (1936, 1937).

Weigall, Lady Rose. *Lady Rose Weigall*, by Rachel Weigall (1923).

Wesley, Elizabeth Ann. *Memorials of Elizabeth Ann Wesley, the Soldiers' Friend*, by the Revd. Samuel Wesley (1887).

West, Mrs John. *A Memoir of Mrs. John West*, by John West (1840).

Westlake, Caroline J. *In Memoriam. Caroline J. Westlake* (Leominster, 1914).

Weston, Agnes. *My Life among the Blue-Jackets* (1909).

—— *Our Blue Jackets. Miss Weston's Life and Work among our Sailors*, by Sophia G. Wintz (1890).

Whately, Mary Louisa. *The Life and Work of Mary Louisa Whately*, by Elizabeth Jane Whately [1890].

Whiteford, Emma. *Emma Whiteford*, by the Revd. Samuel Oliver (1852).

Wigham, Sarah Elizabeth. *Memorial of S. E., daughter of J. and S. Wigham* (Edinburgh, 1855).

Wightman, Mrs J. B. *Mrs. Wightman of Shrewsbury*, by J. M. J. Fletcher (1906).

Wilkinson, Kitty. *Memoir of Kitty Wilkinson of Liverpool, 1786–1860*, ed. by Herbert R. Rathbone (Liverpool, 1927).

Wilkinson, Susanna. *Extracts from the Diary of Mrs. Susanna Wilkinson* (1832).

Wilson, Elizabeth C. *Memoir of a Beloved and Long-Afflicted Sister*, by William Carus Wilson (1842).

Winslow, Mary. *Life in Jesus. A Memoir of Mrs. Mary Winslow*, ed. by Octavius Winslow (1856).

Wolff, Sarah Jane Isabella. *Memoir of Sarah Jane Isabella Wolff*, by the Revd. M. S. Alexander (1840).

Woods, Margaret. *Extracts from the Journal of the late Margaret Woods, from the Year 1771 to 1821* (Philadelphia. 1850).

Yates, Agnes. *Putting the Clock Back. Reminiscences of Childhood in a Quaker Country Home* . . . [1940].

Yates, Elizabeth. *The Benefits of Sunday-School Instruction exemplified. A Memoir of Elizabeth Yates* (1821).

Others: Unidentified Subjects and Group Biographies

Aveling, Thomas William. *Memorials of the Clayton Family* (1867).

Brightwell, Cecilia Lucy. *Memorial Chapters in the Lives of Christian Gentlewomen* [1873].

Brief Memoirs of Remarkable Children, 2 vols. (1823).

Chappell, Jennie. *Four Noble Women and their Work* (1898).

—— *Noble Work by Noble Women* [1900].

—— *Three Brave Women* [1920].

—— *Women of Worth* [1908].

—— *Women who have Worked and Won* [1904].

Crosland, Mrs Newton. *Memorable Women: The Story of their Lives* (1854).

Early Recollections. By Bertie's Mother, ed. by J. Byres Laing [1851].

Fawcett, Millicent Garrett. *Some Eminent Women of our Times* (1889).

Gardner, The Revd. James. *Memoirs of Christian Females* (Edinburgh, 1841).

Hack, Mary Pryor. *Christian Womanhood* (1883).

—— 'Claudia'. *Consecrated Women* (1880).

—— *Faithful Service: Sketches of Christian Women* (1885).

—— *Self-Surrender* (1882).

How, Frederick Douglas. *Noble Women of our Time* (1901).

[J. Johnson]. *Heroines of our Time: Being Sketches of the Lives of Eminent Women* [1860].

Life as I saw It (1924).

Life as We have known It, ed. by Margaret Llewelyn Davies (1977).

Memoirs of Female Labourers in the Missionary Cause, introduced by Richard Knill (Bath and London, 1839).

Memoirs of Six Sisters, ed. by W. Benson (1895).

Memorials for a Wife; dedicated by her Husband to their Children (1856).

Memorials of a Beloved Mother, by M. C. F. [1893].

Pearls from the Deep: Consisting of Remains and Reminiscences of Two Sisters [1852].

Ross, Janet. *Three Generations of Englishwomen. Memoirs and Correspondence of Mrs. John Taylor, Mrs. Sarah Austin, and Lady Duff Gordon*, 2 vols. (1888).

Tabor, Margaret E. *Pioneer Women* (1927).

Tomkinson, E. M. *Sarah Robinson. Agnes Weston. Mrs. Meredith* (1887).

Wakeford, Constance. *The Prisoners' Friends: John Howard, Elizabeth Fry, and Sarah Martin* [1917].

X. Secondary Sources

Annan, N. G. 'The Intellectual Aristocracy'. *Studies in Social History*, ed. J. H. Plumb (1955), pp. 241-87.
Ariès, Philippe. *Centuries of Childhood* (1973).
Banks, J. A. *Prosperity and Parenthood* (1954).
Basch, Françoise. *Relative Creatures: Victorian Women in Society and the Novel, 1837-67* (1974).
Berg, Barbara. *The Remembered Gate: Origins of American Feminism* (New York, 1978).
Best, Geoffrey. *Mid-Victorian Britain, 1851-1875* (1971).
Bradley, Ian. *The Call to Seriousness* (1976).
Bristow, Edward. *Vice and Vigilance* (Dublin, 1978).
Brown, Ford K. *Fathers of the Victorians* (Cambridge, 1961).
Burn, W. L. *The Age of Equipoise* (1964).
Canton, William. *A History of the British and Foreign Bible Society*, 5 vols. (1904-10).
Chadwick, Owen. *The Victorian Church*, 2 Parts (1966, 1970).
Clark-Kennedy, A. E. *Edith Cavell* (1965).
Cook, Sir Edward. *The Life of Florence Nightingale*, 2 vols. (1913).
Cott, Nancy F. *The Bonds of Womanhood, 'Woman's Sphere' in New England, 1780-1835* (New Haven and London, 1977).
Coutts, General Frederick. *No Discharge in this War* (1975).
Coveney, Peter. *The Image of Childhood* (1967).
Deane, Phyllis and Cole, W. A. *British Economic Growth, 1688-1959* (Cambridge, 1962).
Delamont, Sara. 'The Contradictions in Ladies' Education'. *The Nineteenth-Century Woman: Her Cultural and Physical World*, eds. Sara Delamont and Lorna Duffin (1978), pp. 134-63.
Findley, G. G. and Holdsworth, W. W. *The History of the Wesleyan Methodist Missionary Society*, 5 vols. (1922).
Fox, Lionel W. *The English Prison and Borstal Systems* (1952).
Gidney, The Revd. William Thomas. *The History of the London Society for Promoting Christianity amongst the Jews, from 1809 to 1908* (1908).
Gray, B. Kirkman. *A History of English Philanthropy* (1905).
Hall, M. Penelope and Howes, Ismene V. *The Church in Social Work* (1965).
Hammond, J. L. and Barbara. *James Stansfeld. A Victorian Champion of Sex Equality* (1932).
Harrison, Brian. *Drink and the Victorians* (1971).
— 'For Church, Queen, and Family: The Girls' Friendly Society 1874-1920'. *Past and Present*, Number 61 (Nov. 1973), pp. 107-38.
— 'Philanthropy and the Victorians'. *Victorian Studies*, ix (June 1966), pp. 353-74.
— *Separate Spheres: The Opposition to Women's Suffrage in Britain* (1978).

Harrison, Brian. 'State Intervention and Moral Reform in nineteenth-century England'. *Pressure from Without in early Victorian England*, ed. Patricia Hollis (New York, 1974), pp. 289–322.
Heasman, Kathleen. *Evangelicals in Action* (1962).
Hemlow, Joyce. *The History of Fanny Burney* (Oxford, 1958).
[Higgs, Mary Kingsland]. *Mary Higgs of Oldham* [1954].
Hinde, R. S. E. *The British Penal System, 1773–1950* (1951).
Holcombe, Lee. *Victorian Ladies at Work* (Newton Abbot, 1973).
Homan, Walter Joseph. *Children & Quakerism* (Berkeley, 1939).
Inglis, K. S. *Churches and the Working Classes in Victorian England* (London and Toronto, 1963).
Isichei, Elizabeth. *Victorian Quakers* (1970).
Jones, Gareth Stedman. *Outcast London* (Oxford, 1971).
Jordan, W. K. *The Charities of London, 1480–1660* (1960).
—— *Philanthropy in England, 1480–1660* (1959).
Kitson Clark, G. *The Making of Victorian England* (1962).
Laqueur, Thomas Walter. *Religion and Respectability: Sunday Schools and Working Class Culture, 1780–1850* (New Haven and London, 1976).
Lascelles, E. C. P. 'Charity'. *Early Victorian Engand 1830–1865*, ed. G. M. Young, 2 vols. (1934), pp. 317–47.
Lorimer, Douglas A. *Colour, Class and the Victorians* (Leicester, 1978).
McCord, Norman. *The Anti-Corn Law League, 1838–1846* (1958).
McCrone, Kathleen E. 'Feminism and Philanthropy in Victorian England: the Case of Louisa Twining'. Canadian Historical Association, *Historical Papers* (1976), pp. 123–39.
Magnuson, Norris. *Salvation in the Slums: Evangelical Social Work, 1865–1920* (Metuchen, NJ, 1977).
Manton, Jo. *Mary Carpenter and the Children of the Streets* (1976).
—— *Sister Dora. The Life of Dorothy Pattison* (1971).
Mitchell, B. R. *Abstract of British Historical Statistics* (Cambridge, 1971).
Moorhouse, Geoffrey. *The Missionaries* (1973).
Mowat, Charles Loch. *The Charity Organisation Society, 1869–1913* (1961).
North, Eric McCoy. *Early Methodist Philanthropy* (New York, 1914).
Owen, David. *English Philanthropy, 1660–1960* (Cambridge, Mass. and London, 1964).
Pinchbeck, Ivy and Hewitt, Margaret. *Children in English Society*, 2 vols. (1973).
Pinchbeck, Ivy. *Women Workers and the Industrial Revolution, 1750–1850* (1930).
Platt, Elspeth. *The Story of the Ranyard Mission, 1857–1937* (1937).
Price, Millicent, *'Inasmuch as . . .' The Story of Sister Dora of Walsall* (1952).
Pringle, J. C. *Social Work of the London Churches* (1937).

Prochaska, F. K. 'Women in English Philanthropy, 1790–1830'. *International Review of Social History*, xix (1974), pp. 426–45.
Sandall, Robert *et al. The History of the Salvation Army*, 6 vols. (1947–73).
Schupf, Harriet Warm. 'Single Women and Social Reform in Mid-Nineteenth Century England: The Case of Mary Carpenter'. *Victorian Studies*, xvii (March 1974), pp. 301–17.
Semmel, Bernard. *The Methodist Revolution* (1974).
Shiman, Lilian Lewis. 'The Band of Hope Movement: Respectable Recreation for Working-Class Children'. *Victorian Studies*, xvii (September 1973), pp. 49–74.
Simey, Margaret B. *Charitable Effort in Liverpool in the Nineteenth Century* (Liverpool, 1951).
Sklar, Kathryn Kish. *Catherine Beecher: A Study in American Domesticity* (New Haven and London, 1973).
Smith, Cecil Woodham. *Florence Nightingale* (1950).
Stock, Eugene. *The History of the Church Missionary Society*, 4 vols. (1899–1916).
Stocks, Mary. *A Hundred Years of District Nursing* (1960).
Suffer and be Still. ed. Martha Vicinus (Bloomington and London, 1972).
Summers, Anne. 'A Home from Home—Women's Philanthropic Work in the Nineteenth Century'. *Fit Work for Women*, ed. Sandra Burman (1979)., pp. 33–63.
Terrot, Charles. *The Maiden Tribute* (1959).
Unsworth, Madge. *Maiden Tribute* (1949).
Veblen, Thorstein. *Theory of the Leisure Class* (1924).
Victorian Nonconformity. eds. John Briggs and Ian Sellers (1973).
Walkowitz, Judith. 'The Making of an Outcast Group: Prostitutes and Working Women in Nineteenth-Century Plymouth and Southampton'. *A Widening Sphere*, ed. Martha Vicinus (Bloomington and London, 1977), pp. 72–93.
Walkowitz, Judith R. and Daniel J. 'We are not Beasts of the Field: Prostitution and the Poor in Plymouth and Southampton under the Contagious Diseases Acts.' *Clio's Consciousness Raised*, eds. Mary S. Hartman and Lois Banner (New York, 1974), pp. 192–225.
Walton, Ronald G. *Women in Social Work* (1975).
Webb, Sidney and Beatrice. *English Prisons under Local Government* (1922).
Woodroofe, Kathleen. *From Charity to Social Work* (1962).
Young, A. F. and Ashton, E. T. *British Social Work in the Nineteenth Century* (1956).
Young, G. M. *Victorian England. Portrait of an Age* (1936).

Index